Engineering Systems Acquisition and Support

Woodhead Publishing Series in Mechanical Engineering

Engineering Systems Acquisition and Support

Edited by

John P.T. Mo, Cees Bil and Arvind Sinha

ELSEVIER

AMSTERDAM • BOSTON • CAMBRIDGE • HEIDELBERG
LONDON • NEW YORK • OXFORD • PARIS • SAN DIEGO
SAN FRANCISCO • SINGAPORE • SYDNEY • TOKYO
Woodhead Publishing is an imprint of Elsevier

WP WOODHEAD PUBLISHING

Woodhead Publishing is an imprint of Elsevier
80 High Street, Sawston, Cambridge, CB22 3HJ, UK
225 Wyman Street, Waltham, MA 02451, USA
Langford Lane, Kidlington, OX5 1GB, UK

British Library Cataloguing-in-Publication Data
A catalogue record for this book is available from the British Library

Library of Congress Cataloging-in-Publication Data
A catalog record for this book is available from the Library of Congress

ISBN: 978-0-85709-212-0

For information on all Woodhead Publishing publications
visit our website at http://store.elsevier.com/

Printed in the United States of America

Contents

List of tables and figures

Biography

Dr. John Mo is professor of manufacturing engineering and former discipline head of manufacturing and materials, School of Aerospace, Mechanical and Manufacturing Engineering at RMIT University. Prior to joining RMIT, he was senior principal research scientist in CSIRO and led several teams including manufacturing systems and infrastructure network systems in the Division of Manufacturing and Infrastructure Technology. His expertise includes system integration, data communication, sensing and signal diagnostics. In his 11 years in CSIRO, he led a team of professional research staff; worked on risks analysis algorithms, electricity market simulation, wireless communication, fault detection and production scheduling. He was the project leader for many large-scale government projects including productivity improvement in furnishing industry and consumer goods supply-chain integration.

Dr. Cees Bil is associate professor of aerospace design at RMIT University in Melbourne, Australia. He has a PhD in Aerospace Engineering, a master of science and a graduate diploma (Education) from Delft University of Technology, Delft, the Netherlands. His research interests include aerospace systems design, through-life support, unmanned air systems, occupant safety, air-traffic management, knowledge-based engineering, feature recognition and dynamics and control. He is currently involved in occupant survivability assessment, unconventional aircraft concepts and autonomous systems.

Dr. Arvind Sinha obtained his PhD from RMIT University, in Aerospace Engineering. He is director of engineering, Defence Materiel Organization in Australia. Prior to working with the Australian Government, he was associate professor in Aerospace Engineering, and director of Sir Lawrence Wackett Centre for Aerospace Design at RMIT. He was also the President of the Australian chapter of the American Helicopter Society (AHS) and pioneered the helicopter education in design and airworthiness at university level. He received two awards from the AHS in 2004 and 2007 recognising his contribution to aerospace engineering.

Preface

The availability of effective global communication facilities nowadays has changed the business goals and practices of many enterprises that are developers of complex engineering products such as an aircraft or a frigate. They need to remain competitive by developing innovative products and processes that are specific to individual customer's requirements, completely packaged and made available globally to make best use of resources within defined constraints. The complexity of these engineering products also means the need for developing a different set of knowledge that includes the following:

- Systems design, development, validation, testing, installation and commissioning during acquisition phases;
- On-going operations support, mid-life upgrades, services and logistics, maintenance and repairs during operations phase;

In some cases, due to regulatory and institutional requirements, consideration is also required for decommissioning.

In the acquisition phase, user requirements are not normally clear. The design of systems can vary greatly between different vendors and concepts. In many cases, several consortia are involved, each bringing their own expertise and expectations. The challenge to the system acquisition team is to establish suitable baseline for comparing different proposals. In order to make a fair judgement to different designs, the acquisition needs to apply modelling methods and analysis tools to assist in the evaluation and interrogating processes.

The lengthy and complicated acquisition process is further complicated by the increasing expectation that the complex engineering system needs to be in service for years. From the owner's point of view, asset capability should increase over time to meet changing demands. To achieve this goal, typical support and services after acquisition of the asset should include significant mid-life upgrade and whole of system integrated logistics support. The philosophy of whole of life engineering development and support is to integrate product-design information through to implementation, manufacture, services and decommissioning of the product. Substantial planning is required to enable flexible delivery of services and logistics support to the asset from anywhere in the world.

To address these issues, business systems across company boundaries are formed to operate across time and space with world-wide distributed functions of systems design, testing, validation, manufacturing, sales, customer support, engineering, quality assurance, project management, supply and process chain management, operations control and so on. New business models such as performance-based contracting has emerged in recent years as one of the favourable choices of contracting mechanisms for the public sector and asset-intensive industries such as defence, water, transport and chemicals. Under performance-based contracting, a third-party contractor takes responsibility and risks for the management of a specific part of the business. This compares to traditional maintenance-oriented arrangement where the asset's capability will inevitably degrade due to age and technology changes.

However, no two systems acquisition and support projects are the same because each system needs to meet different user requirements. Due to the highly individualised nature of the acquisition and support of complex engineering systems, it is an extremely knowledge intensive, labour-rich business that requires the development of new systems in the form of design certification, digital manufacturing, integrated testing, customer-support centres, information infrastructure, human organisations, multi-disciplinary engineering teams, supply chains and distribution networks. Operations of these engineering units in a globally distributed environment present significant challenges to the engineering systems developers.

This book will address the skill set of professional engineers and operations managers required to acquire and support a complex engineering system that is expected to last for a long period of time and to meet increasing demand in capability. This book provides a core knowledge base that draws upon principles derived from a wide range of business and engineering disciplines including systems engineering, services engineering, operations support, integrated logistics, industrial engineering and many others.

The book is intended to be both a special text for relevant courses at university level as well as a general knowledge text book for professional engineers, systems managers, acquisition officers and operations managers to enhance their knowledge in acquisition and support of complex engineering systems. The book is also written with the aim that research students studying in systems engineering, support and services engineering, logistics and procurement will find it useful as an introductory text to start their research aspiration. The broad range of knowledge requirement in acquisition and support demands a broad coverage of both engineering and business perspectives of expensive complex engineering systems. Books in engineering services tend to focus on people management issues, which are not sufficient in diagnosing and managing problems operating engineering systems. This book will:

- Create a new understanding and operational view of modern acquisition, design, services and support systems. Contemporary books on asset management only focus on maintenance and reliability issues that are restricted in scope.
- Include enterprise modelling and analysis techniques that are essential for engineering professionals involved in the acquisition and support of complex engineering systems to develop a whole of systems view of the design and services requirements of such a system in its serviceable life.
- Take a systems-engineering approach that guides the reader through important systems-development processes of systems design, testing and verification for both complex engineering system and post-acquisition support system. This approach is different from many engineering management books that only focus on management of the engineering process rather than the process itself.
- Provide a detailed account of performance and risks-analysis methods that are critical to the acquisition of a complex engineering system and subsequently its satisfactory operation through the life of the system. Books in maintenance and reliability engineering do not cover these analyses at the acquisition phase.

The book uses case studies and work examples to illustrate the concepts and principles. It will be organised in the chapters that follow further.

1. Introduction
   This chapter introduces the background and concepts of acquiring, operating, supporting and servicing complex engineering systems. In particular, the significance of system integration and whole of life systems design.
2. The life cycles of complex engineering systems
   This chapter outlines stages in the life cycles of complex engineering systems and their significance in designing and management. Along with the description of the life cycles, relevant tools and methods for assisting management of the system are explained.
3. Systems acquisition principles
   This chapter explores principles and processes of acquisition of systems. Complex engineering systems are by nature specially designed and often one-of-a-kind applications. Governance of the processes in systems acquisition is maintained though formal acquisition processes such as tenders and design participation.
4. Systems design
   The systems-design process is standard and structured. This chapter explains systems modelling, sub-systems design, tooling and implementation.
5. Management of engineering-design changes
   Engineering changes have the effect of increasing system costs. This chapter discusses the requirements for engineering design change, change impact analysis and project management issues.
6. Systems prototyping and testing
   This chapter covers the testing and certification phases of the development of complex engineering systems. These include methodologies and processes in prototyping, digital manufacturing, physical and virtual testing, design verification and certification.
7. Operations support and logistics
   This chapter covers the system design and operation of support systems using different theoretical tools including process mapping, operations research, logistics systems analysis and remote support.
8. Systems maintenance and reliability
   Maintenance and reliability are concepts that are closely linked together. There are many types of maintenance, each with special characteristics. This chapter covers the statistical principles and their application to maintainability and reliability assessment.
9. Mid-life upgrades
   The capability of complex engineering systems requires constant review and upgrade. This chapter covers analysis of systems' upgradability, the issues of data authority and certification. Approaches such as the blackbox approach for mid-life upgrade projects are discussed.
10. Architecture for designing support systems
    This chapter explores the architecture of support systems that provides the basis for new effective system design. The chapter presents enterprise modelling methodologies and systems architecture concepts and discusses the support systems design and models' evaluation requirements.
11. Performance and risks analysis
    The chapter covers the means and trends that can be developed as key performance indicators for assessing risks in the operation of new support systems. Common methods such as failure mode and effect analysis are explained.

12. Continuous improvement
Continuous improvement is the key to maintaining a sustainable engineering system, both in terms of its reliability as well as continuity of use. The essential cycle of plan–do–check–act is discussed with examples.

John Mo
Cees Bil
Arvind Sinha
May 2014

Introduction

1

In Globalization 1.0, which began around 1492, the world went from size large to size medium. In Globalization 2.0, the era that introduced us to multinational companies, it went from size medium to size small. And then around 2000 came Globalization 3.0, in which the world went from being small to tiny.

Thomas Friedman

Chapter Outline

1.1 A new business environment for complex engineering systems

Physically, the world does not shrink. What actually changes is the ability of people to interact with one another, directly among themselves or indirectly through some mechanisms or processes; the latter have doubled or tripled in a short time. Developers of complex engineering products such as aircraft or frigates are facing new challenges in meeting business goals and competition globally. They need to remain competitive by developing innovative products and processes which are specific to individual customers' requirements, are completely packaged and are made available globally to make best use of resources within defined constraints. New business models such as performance-based contracting have emerged in recent years as some of the favourable choices of contracting mechanisms for the public sector and asset-intensive industries such as defence, water, transport and chemicals. Under performance-based contracting, a third-party contractor takes the responsibility and assumes the risks for the management of a specific part of the business. This

Engineering Systems Acquisition and Support

approach compares to traditional maintenance-oriented arrangements in which the asset's capability will inevitably degrade due to age and technology changes.

The complexity of these engineering products also means the need to develop a different set of knowledge that includes:

- Systems design, development, validation, testing, installation and commissioning during acquisition phases.
- Ongoing operations support, mid-life upgrades, services and logistics, maintenance and repairs during the operations phase.

In some cases, due to regulatory and institutional requirements, decommissioning must also be considered. Thus, this new knowledge set is necessary for professional engineers and operations managers working in the new business environment so that they may acquire and support complex engineering systems that are expected to last for a long period of time and to meet increasing demands in capability. To examine this core knowledge base, we'll use a number of examples to illustrate the key elements.

1.2 Examples of complex engineering systems

A characteristic that is common to all complex engineering systems is the complexity and multidisciplinary coordination of the support system that enables the smooth operation of the engineering system.

1.2.1 Metro Trains Melbourne (MTM)

Metro Trains Melbourne (MTM) is a consortium of rail and construction businesses including Hong Kong's MTR Corporation, Australia's John Holland Group and UGL Rail, a division of United Group Limited. MTM has a contract to operate and maintain the public transport rail service used by Melbourne commuters.

This franchise contract brings together international experience to deliver long-term service-supply agreements that cover rolling stock supply, maintenance and refurbishment, rail infrastructure maintenance and renewals, and the delivery of major capital works programme. MTM was selected as the new operator by the State Government of Victoria in June 2009 and replaced the previous operator, Connex Melbourne, on 30 November 2009.

MTM operates a fleet of approximately 323 trains on 830 km of track. There are 16 daily train lines and 1 special-events train line. The train fleet covers more than 30M km per year and services more than 200M customer journeys per year. MTM is also responsible for 211 train stations and employs a workforce of 3,500 rail professionals that includes train drivers, mechanical and electrical engineers, network operations specialists and customer-service representatives. As part of the agreed-upon service plan, the Victorian Government (owner of Melbourne train network) promised to increase the carrying capacity by approximately 20%.

MTM's operations and maintenance activities are governed by legislation, regulations and guidelines including:

- Transport Integration Act 2010
- Rail Management Act 1996
- Rail Safety Act 2006
- Transport (Compliance and Miscellaneous) Act 1983
- Transport (Ticketing) Regulations 2006
- Transport (Conduct) Regulations 2005
- Victorian Fares and Ticketing Manual

A key part of the operator service contract is the achievement of high punctuality (Metro Trains, 2013). Metro had a good record for services run; the percentage exceeded the target of 98% from November 2009 through June 2010 (i.e. every one of the first eight months after the consortium took control of the network). However, on 27 July 2010, a major service disruption was recorded that affected 400,000 passengers. In addition, according to the overall performance measure, MTM incurred performance penalties of $4.8M from December 2009 to June 2010. Trains were mostly delayed due to electrical fault in the system and operator delays.

Analysis of the initial MTM operation and maintenance model shows the following deficiencies:

1. *Customer satisfaction.* The Public Transport Performance Report for the six months ending December 2012 indicated a customer satisfaction index score of 68%. Although MTM came close with a 67.3% score, this level was far from the 90% that the London Overground achieved in the preceding period.
2. *Service punctuality.* The same report lists the service punctuality at 92.4%, which exceeds the 89% specified target. However, the Hong Kong rail system carries over 4M people a day and manages to achieve 99.9% on-time arrivals.
3. *Inefficient operations.* Before 2011, train timetables had not been redesigned holistically since the City Loop was opened in the 1980s. Rather, additional services were generally added into the timetable incrementally, often by only changing the times of adjacent trains to accommodate new services. This approach resulted in the complex, irregular and inefficient stopping patterns and service headways that currently remain on some lines.
4. *Maintenance.* As a direct consequence of previous franchise that simply maintained the status quo of assets and operational expenses instead of planning and implementing an improvement regime, MTM has been left with a severely degraded asset. MTM's asset management plan (AMP) is reported to portray a high level of reactive maintenance compared to planned maintenance, which is to be expected given the state of the inherited infrastructure.

Influences on these deficiencies come from a myriad of sources. Internally, MTM has the desire to improve the reliability of the infrastructure to permit a higher level of reliability and therefore improved train service reliability. This strategy means that, in the short term at least, there can be expectations of higher-than-desired levels of impact on service delivery due to the high levels of maintenance and upgrades that are required in order to meet those long-term goals.

Other deficiencies in urban railways are often influenced by the environment in which the system is being operated. Like many Western countries, Australia has an excessive dependence on automobiles which has led to underutilised public-transport systems. This underutilisation has resulted in poor-performing services as the costs of maintaining the assets and infrastructure outstrip income for the organisation or government operating the transport solution. The flow-on effect is reduced spending on upgrades and new infrastructure and a lack of investment in the industry.

One thing that appears absent from consideration is how to make public transport look attractive to commuters who currently use private transport. Key to this goal is a reliable, consistent service that is safe, quick and affordable. The way to achieve these improvements is through collaboration throughout the supply chain in ways that ensure the interfaces between logistics and maintenance are efficient and effective and ensure that services are supplied on time and with sufficient capacity. In addition, rationalising the road network and placing public transport as a higher priority than vehicle access could encourage greater use of public transport and reduce road-maintenance costs.

1.2.2 Offshore wind farm

The London Array offshore wind farm comprises 175 turbines (3.6 MW Siemens turbines) that generate 630 MW of power with more than 450 km of cabling connecting them to two offshore substations and one onshore substation. The turbine hub height above sea level is 87 m with a 120 m rotor diameter. The wind farm is located 46 km off the coast of Kent and Essex in the UK, at an average water depth of 20 m.

Company revenues are based directly on the energy provided to the electricity grid on a $/kWh. Therefore, the operator of the wind farm is interested in 24 hours a day, 7 days a week operation that maximises the total energy produced by each wind turbine. The total energy produced by a wind turbine is a product of the prevailing wind speed, the power output curve of the wind turbine and its availability (Karyotakis, 2011).

A wind turbine is a complex collection of components comprising rotors, pitch system, drive train, gearbox, generator, electrical system, mechanical brakes, yaw system, sensors, control system and hydraulics. The wind-turbine farm also consists of towers, foundations, underwater power, communications cables and substations. To support a wind farm, four maintenance categories have been defined:

- Replacement of rotors, nacelles, gearboxes and generators with external cranes.
- Replacement of large components with internal crane.
- Replacement of small parts.
- Inspection and repair.

A centralised maintenance facility located at the Port of Ramsgate is used to accommodate 90 staff and 6 maintenance vessels, workshops, spare-part storage and a computerised control and monitoring system. The estimated lifetime of each wind turbine is 20 years.

Based on the four maintenance categories, regular service visits are made to wind farms at (generally) six-month intervals with major lifting operations performed every 3–5 years to perform overhaul of the turbines. The London Array comprises 175 turbines, which would result in 30–35 lifting actions per year. Given the 87 m hub height of the turbines, such actions require a very large stabilised crane ship to perform this maintenance. It appears that a crane ship is not part of the maintenance fleet of six ships and so would need to be hired at great cost and uncertain availability.

A number of periodic maintenance and inspection tasks would also need to be carried out. These include oil and hydraulic fluid-level checks, environmental monitoring equipment checks, cable checks, foundation-movement checks and substation checks. Given the sophistication of the remote health-monitoring system, a large part of this work will be done from the Ramsgate facility without anyone travelling to the site.

The London Array wind-farm-support solution has a single level of onshore maintenance located at the Port of Ramsgate maintenance facility. There are no operational or intermediate levels of maintenance due to the remoteness of the wind farm and the difficulties of weather and sea state. The Ramsgate facility provides storage of the spare parts and consumables as well as workshops to carry out repairs. The six maintenance vessels that are used to transport personnel and equipment to the turbines are also located at this facility. Design of the support system can be considered in relation to the criticality of system outage. Unlike an oil rig that must maintain high reliability to reduce high-cost outages, the wind-farm operation can take a more risk-based approach as the high number of wind turbines provides excess capacity for individual turbine failures to occur and for the corrective maintenance to be delayed until it is more convenient to perform it without impacting the performance targets. A reactionary, highly interventionist maintenance strategy will result in high maintenance costs due to the large travel requirements for equipment and personnel to the remote wind farm. High costs are not tolerated in the highly competitive market for electricity generation and supply.

The ability to initiate coordinated service actions will require a sophisticated computerised maintenance management system (CMMS) at its core. The CMMS will require access to large amounts of data from all facets of the wind farm's operational and maintenance aspects. Some of the considerations for the CMMS for the London Array are:

- Part-reliability data.
- Condition-based monitoring (CBM) data.
- Environmental data.
- Spare-parts data.
- Maintenance schedule data.
- Maintenance personnel and equipment.
- Energy generation/supply data.

The part-reliability data will comprise extensive lists of MTBF information for system components that are kept up to date with data from the CBM (run times, temperatures, failures) and environmental (wind speeds, wave heights, rain) functions of the CMMS. Keeping the lists up to date will allow maintenance schedules to be reviewed and optimised accordingly.

A CBM strategy is implemented on the London Array through the Port of Ramsgate remote monitoring-and-control system. Each wind turbine is connected on a network that allows health monitoring information to be gathered. The control function also allows adjustments to the wind farm in response to reported changes/faults.

A key application-software package that would retrieve data from the CMMS and facilitate maintenance management would be a wind-farm maintenance-task scheduler (WFMTS). As stated previously, redundancy is built into the design of the London Array in the form of extra wind turbines that are surplus to the baseline power-generation performance targets. This redundancy allows corrective maintenance to be delayed and aligned with some other key maintenance cost drivers such as other planned maintenance activities, accessibility to turbines and specialised equipment, time limitations and power generation quota. Note that occupational health and safety (OH&S) does not allow work to be conducted on wind turbines in locations with wave heights greater than 2 m, which is mostly the case for the London Array between November and April.

A planned corrective maintenance strategy for offshore wind-farm maintenance has enormous benefits for the optimisation of availability and reduction of life-cycle costs. Although not much can be done to improve the inherent turbine reliability, much can be done to decrease the maintenance down-time by improving asset accessibility as follows:

- Corrective maintenance can be aligned with the planned maintenance activities of other turbines.
- Planning means that expensive and hard-to-get assets such as ship cranes can be pre-booked at convenient times.
- Work can be scheduled within optimum maintenance windows (e.g. favourable weather and sea state conditions).
- Spare parts can be provided in a just-in-time fashion, which will improve response times and reduce inventories and holding costs.

1.2.3 Railway rolling stock and network

In a long-distance railway system, rolling stock comprises all the vehicles that move along a railway, including powered and unpowered vehicles such as locomotives and passenger carriages. In mass-transit networks, the station distances are short which leads to short braking and acceleration cycles and increased mechanical wear of components. Failures of operational rolling stock due to poor maintenance practices causes delays and lost revenues for the operator.

Rolling stock maintenance costs are 30% of the total life-cycle costs of the asset; 60% of these costs are related to personnel and the remaining 10% to spare parts (Higgins et al., 1999).

Some typical maintenance activities performed on rolling stock are:

- Lubrication checks: accessible linkages and moving parts are oiled.
- Examinations/Inspections: safety and wear checks are made regularly to inspect condition of asset.
- Minor overhauls: internal and external cleaning.
- Major overhauls: repairs of bogies and diesel and electrical traction motors, and wheel grinding.

The support system for rolling stock is complicated by wide variations in the ages of vehicles and the various different configurations of carriages. Thus, a large amount of coordination is required to ensure that the right maintenance equipment, spare parts and qualified personnel are available at the right locations at the right times.

A rolling stock support system has three levels of maintenance: operational, light and heavy. The operational-level maintenance of the rolling stock might include lubrication and inspection of components during refuelling or replacement of light bulbs or other consumables at convenient times. The spare parts and consumables would be kept at stations for ready use. The light-maintenance work will be performed in a light maintenance depot (LMD) during time slots when no train service is required (i.e. night time) and will include minor overhaul work of short duration that doesn't remove rolling stock from operational availability. The heavy maintenance depot (HMD) will house major overhaul work that makes the rolling stock unavailable for operational use for long periods of time.

The rolling-stock brakes and wheels are high-wear items due to the stress of constant rapid accelerations and stopping along the train route. This problem is worse for trains in the denser inner-rail networks (where there is more start-stop) compared to the less-dense outer-rail networks. To restore the rolling stock to working order, wheel grinding and/or change of brakes must occur. These activities require heavy maintenance that makes the rolling stock unavailable for long periods of time and means that a standby train must be utilised until this work can be performed.

In conjunction with the rolling stock, a railway network is a fixed-infrastructure asset that comprises the following major items:

- Track and track bed.
- Safety, signalling and telecommunications installations.
- Bridges, tunnels and culverts.
- Level crossings.
- Plant(s) for transforming and carrying electrical power.

The safety and signalling systems are usually a combination of old mechanical systems and new electronic systems that require a diverse range of parts and training skills for maintenance. The train-signalling system exists for the safe operation of the railway and provides a fail-safe interlocking system to lock out trains.

Due to the fixed nature of the railway infrastructure, the majority of services are performed *in situ*. Some of the rail-maintenance activities are:

- Rail grinding, which is conducted to correct rail corrugations, fatigue and metal flow and to re-profile the rail.
- Rail replacement due to defects, wear or derailment damage.
- Tamping, which is conducted to correct the longitudinal profile, cross-level and alignment of track.
- Track stabilisation, which restores the lateral resistance of the rail to the original level.
- Stoneblowing, which adds stones to the ballast bed to correct longitudinal rail profile.
- Sleeper replacement, which replaces defective rail sleepers.

These activities are a mix of scheduled and corrective actions that must be done by the support system. Unlike wind turbines, the preventative maintenance carried out on railway infrastructure is related directly to safety matters. Failures cannot be allowed to happen due to the catastrophic consequences for the asset and for people, whereas corrective maintenance has more to do with passenger comfort. In optimising the availability of the rail infrastructure, the maintenance work (planned and unplanned) must occur when the trains are not operationally using the railway. This scheduling requires careful coordination between the infrastructure operator and the train operator to ensure there is no overlap that will delay maintenance and increase the downtime.

Maintenance tasks are prioritised. Rail grinding, tamping, ballast profiling and screening are performed by specialised machines that allow personnel to travel over the rail network at high speeds. More general-purpose kinds of equipment are used for other general repair work. Daily patrols and inspections of infrastructure for signs of problems are conducted especially after high-rainfall events that can compromise rail network integrity and signalling systems.

Railway availability and costs are also affected by the limited availability of accommodation for maintenance personnel in remote areas due to high demand from the mining industry. In many cases staff must travel longer distances for work, which increases their fatigue and reduces the time for them to do actual maintenance work.

1.3 Value for money

Due to the high value of complex-engineering assets, the acquisition process is governed by a set of procurement procedures that ensure fairness and probity. The purpose of acquisition is to increase the capability of the operating system over a defined period of time. Government acquisitions are usually massive due to the systems involved, such as infrastructures and defence. The primary objective of all government procurement is to obtain value for money (VfM). This means obtaining the best possible outcome after taking into account all relevant costs and benefits over the life of a procurement.

Modern governments have substantially increased the proportion of sustainment services outsourced to engineering-system suppliers in a drive for improved operational effectiveness at reduced cost. They have also taken a more-strategic approach to outsourcing by asking industry to provide solutions for complete capability, as opposed to piecemeal support packages. For example, the UK Ministry of Defence (MoD) is moving from support provider to intelligent decider by requiring contractors to join with them in taking a through-life-capability management approach (Ministry of Defence, 2005). In the United States, the Department of Defence is increasingly moving towards performance-based logistics. The Australian Department of Defence has adopted availability contracting and is targeting reduced costs of ownership. These changes are driving a different business approach by industry, just as they reflect a different business approach by defence customers.

Modern economies are characterised by massive infrastructure investment to provide fundamental community services and to facilitate economic growth. The more advanced the economy is, the greater the level of infrastructure investment can be. The infrastructure assets include energy supply and distribution, water supply and distribution, telecommunications, transportation systems (including ports and airports) and local community assets such as sporting facilities, as well as industrial infrastructure. Such infrastructures may be owned by the public sector or the private sector. There is a trend towards increasing private-sector ownership of infrastructure assets, under the theory that private investors have the drive to maximise the VfM of their assets.

In the drive to achieve VfM, the shift to performance-based or output-based contracting means moving the risk of operating the asset to the private sector by changing contractor roles and increasing contractor responsibilities during the acquisition process. The increasing share of risk associated with the support of complex assets makes the design of the business solution as challenging as that of the technical solution.

A performance-based contract will not produce ongoing reductions in cost by itself. Instead, achieving performance outcomes and sharing in ongoing productivity gains and other cost-reduction opportunities require cooperation of compromise among all parties concerned. Reductions in cost often originate from efficiency gains implemented by the contractor and from productivity improvements that span asset owner and contractor organisations. They also result from shifting responsibilities from asset owner to contractor due to economy of scale, and from shifting work from the asset owner to the contractor to reduce costs by consolidating common businesses. Some of the resultant savings are passed on to asset owner through a reduction in the contract price.

A performance-based contract, based on a firm-priced contract, already offers the contractor greater profit if it can enable asset owner to achieve its desired outcomes more efficiently within the contract price. However, such a contract does not provide any mechanisms to enable the asset owner to share in those savings. To make a performance-based contract effective, additional provisions must be included that enable the asset owner to share in the savings from productivity improvements. One way to make this happen is to design a reward when savings over the current term are achieved and the required level of performance is maintained. Larger-support contracts may use other, more complex methods, such as a target-cost incentive model. Balancing the savings and the performance needed to achieve asset owner's desired outcomes can involve complex interactions among the different mechanisms, their costs and the resulting outcomes.

1.4 Requirements of logistics for support

A support system has many functional activities (Johansson and Olhager, 2006). Maintenance is the only activity in the support system that literally touches the assets; thus, it has a unique position in the value-delivery chain. Whether one is maintaining the approved configuration of the asset system or embodying engineering changes within that system, the maintenance function is the primary activity of the business solution that physically affects the asset. The maintenance function communicates to the asset operators their maintenance requirements in terms of timing, extent, capacity, regulatory requirements and other constraints as the maintenance work is performed. Correspondingly, the operators need to communicate to the maintenance organisation any special maintenance requirements not already included in standard scheduled servicing.

To support maintenance, supplies and repairable items are managed through the supply-support function. Although some supplies are provided directly to the operating organisation for routine daily maintenance, the bulk are used or consumed on the assets within the major scheduled maintenance programs.

In its broadest context, logistics encapsulates the functions that are required in the planning, operation, support and maintenance of an asset/fleet. In essence, logistics is the bridge between the asset/fleet and the industrial base that produces the material and equipment needed to keep the asset/fleet in operation. Crossing this bridge can involve the identification of requirements, as well as the building up of stocks and capabilities to sustain the asset/fleet.

The supply-support function employs appropriate management methods, practices and procedures to determine the exact requirements for goods and services as well as their acquisition, receipt, storage and issuance. Materials required for maintenance can include spare parts, labour, tools and facilities. Information on approved spare parts, labour and tools is communicated to the supply-support department, where suppliers are identified and by which the materials are procured in a timely fashion to ensure that the maintenance can be done within the specified time frame. Critical to the maintenance function is the feedback on stock lead-times and the availability of facilities, tools and labour to assist in maintenance planning.

Support engineering involves the logistics and management of supplies on a global basis. The latest e-commerce and e-service business models have extended manufacturing activities of complex-system engineering companies beyond their traditional business environment. By using new developments in information technologies for supporting information infrastructure systems, the expertise employed in manufacturing of a complex system can be deployed to support that system during operations.

As systems first become operational, a number of problems are likely to occur whose solutions require procedural adjustments and modifications. Experience has indicated that many such problems, particularly for large systems, are related to inadequacies of product support and fall into the following categories:

- Individual elements of logistics are incompatible with the prime elements of the system. For example, the prescribed test equipment will not perform the proper functions in verifying prime system operation or in the performance of maintenance.
- The depth and extent of support provided are insufficient. For example, not enough of the right type of spare parts are available; the personnel assigned to operate and maintain the equipment are inadequately trained for the job; and so on.
- The level of support is greater than what is actually required. For example, facilities, personnel and support equipment are not being fully utilised or there are too many spare parts of a certain type. Such situations result in a higher inventory cost than necessary.
- The elements of logistic support are not compatible with each other. For example, the maintenance procedures cover neither the tasks being performed at a given level nor the support equipment used in task accomplishment.

Appropriate asset availability requires both advance planning and procurement of the necessary spare parts and consumables. In general, the larger the activity, the earlier the planning should commence. A change in available service should be accommodated in the pre-planned work packs, schedule, parts and personnel to ensure that the change can be fitted within the availability.

For example, in the context of naval assets, the integrated logistics support (ILS) department creates or updates maintenance routines based on an approved maintenance plan. After taking advice from experts in areas that include logistic engineering, design and engineering, and operations, the support-services company will conduct the maintenance as tasked by the derived contract. Stock holdings are recommended in the form of a spares-assessment report that covers what quantities are to be held at each physical location and storage facility.

Stock-holding levels are reviewed from time to time based on the demonstrated reliability performance that affects consumption due to unplanned maintenance requirements and upon the preventive maintenance plan. Stock-holding levels are also reviewed according to the lead time that is required to refurbish items that have significant impact upon the maintenance schedule, to ensure the availability of these items.

Some support services can be equipped with a standard spares kit. Such kits may contain many items, but maintenance work in the field may require something that is not included. Most kits contain seals on replacement parts that are to be used only as required; usually such seals are on limited-life items that are accompanied by objective quality evidence to define their fitness for purpose. Once a kit has been opened, all configurations, protective seals etc. are lost. Although it is not unheard of for people to have a ready supply of kits, keeping inventory of kits in the field can pose challenges in terms of space and suitability.

1.5 Lean support services

The support-services concept involves the integration of service with system engineering to design support solutions (Dibia et al., 2012). It incorporates a core knowledge base that has to be amalgamated with technical knowledge that is specific to the asset being supported (i.e. the on-asset technologies). The support system will be put into service. Service on assets that are essential to the key outcome of a support system is qualitatively different from the familiar product-based approach in which hard artefacts are delivered to the asset owner. Service is a negotiated exchange with the asset owner (and operator) to provide intangible outputs that are usually co-produced with the asset owner and consumed proximately to the time of production. This means that services cannot be transferred to other asset owners in the same way that products can. The system support engineer therefore requires service knowledge to anticipate management of the support system during implementation.

The successful management of sustainment activities is critical to contractors and customers alike. The management of support risks and opportunities, process inputs, outputs and dependencies must be fully understood to ensure that the support solution meets customer requirements while also offering a continuously improving value proposition. In service-and-support solutions, technical and engineering elements abut with business elements such as supply chain and operations management in such a way that all elements are fully integrated in service delivery to reduce or eliminate waste.

In this respect, classic management theory is insufficient for complex engineering systems based on maintenance-focused processes such as performance monitoring, process control and fault-diagnosis techniques. Many factors (e.g. opportunity costs and satisfactory mission fulfilment) are difficult to quantify and measure. Many decisions about asset acquisition and support are therefore made according to rules of thumb rather than analysed system-performance data. Many contingency allowances are built in during the initial design of the support system. For example:

- An extra standby generator that is idle most of the time. Idleness is a waste of the utilisation opportunity of the asset.
- Flight service not filled to maximum capacity. Empty seats in flight or transport equipment are rotten product.
- Maintenance team and equipment waiting for an operation to stop in order to perform a maintenance task. Waiting is time wasted for all other assets or systems that should be working.
- Contingency costs due to redundancy. Redundant materials are often unused and end up being destroyed.

The above are the primary candidates for waste elimination by engineering design and system improvement. Note that optimisation decisions should be based on the proviso that no critical setup has been ignored for the purposes of cost reduction.

Agile service is a concept that is complementary to lean support services; it addresses the demand for more-flexible support outcomes as the enterprise becomes lean. Agility is usually more capital-intensive than lean support because it involves a high degree of information-technology development. Unfortunately, however, resources allocated to agility projects are always limited. Management should therefore maximise the effectiveness of these investments. To date, very little research has been done on the principles of how agile service practices should be implemented with the support of IT. By contrast, the performance of lean programs depends on workplace practices that are both socially and technically oriented. There are no clear guidelines on how leanness and agility can be implemented in combination, even though combining them has been shown to be more effective.

In short, integrating lean and agile characteristics can achieve better work flow. Moreover, it is clear that productivity-improvement projects, which are linked to some kind of investment in capital, resource usage, commitment, opportunity loss etc. should include the application of both lean and agile service concepts. The question of how the benefits of a combination of lean and agile strategies can be maximised in a service environment has yet to be answered.

1.6 Concept of integration

The need for integration in support engineering is fundamental to the nature of a support system. Multiple partners with incompatible systems will have to work together to produce the final outcome. To provide the best solution to the customer, many manufacturing, industrial, service and commercial activities are grouped into collaborative teams within networked organisations. The operating conditions of the business environment are characterised by frequent changes in products, services, processes, organisations, markets, supply and distribution networks.

An example is the ANZAC Ship Alliance. The ANZAC-class frigate is designed to meet the requirements of the Australian and New Zealand defence forces (Mo et al., 2006). The first ship was delivered in 1996 after 7 years of product development. The ships are planned to have a service life of 25–30 years. In this time period, there will be changes in the operational requirements for these vessels. Major changes include upgrades of the mission systems and platform systems, as well as operational changes. The long project time means that people, skills, experiences, companies, society and government will change. However, the ship-building industry operates in a project-based management structure. Subcontractors and contract staff are employed to fulfil the needs of the project at different times; in addition, the ship-building team can change frequently. Information must be made available in a consistent fashion to people who are joining the project at different times.

The ANZAC Ship Alliance (ASA) is an organisation formed by an agreement between the ship builder, the system OEM supplier and the customer with the aim of providing design and implementation of changes to in-service ANZAC-class frigates over the lifetime of the product (i.e. the frigates). This organisation is an attempt of all parties involved in the creation of the ANZAC fleet to provide a stable support environment that ensures the ships are combat ready at any given time with the best value outcome. The alliance is not a legal company entity. It draws its resources (e.g. staff and assets) from the member organisations as required. It is essentially a dedicated supply chain that provides fast and responsive changes to the ANZAC ships.

The ASA forms project teams that deliver projects and dissolve when the job is completed. The teams work together as a single entity for a goal but the relationships among them and their parent companies often rely on trust and industry practices. This type of temporary alliance is commonly known as a virtual enterprise. Successful achievement of the goal of a virtual enterprise demands well-coordinated agility in all internal and external aspects.

An important aspect of quickly forming customised virtual enterprises is alignment of the objectives of the amalgamated inter-enterprise business processes. The information system is the link that enables people in the virtual enterprise to make wise decisions. Without the right information at the right time, delays and more seriously costly mistakes can occur. The study of the logistics of virtual enterprise is about managing the information flow and data architecture across company boundaries to support the work in the virtual enterprise.

The primary cause of non-integration in this type of service environment is the disjoint of information (Figure 1.6-1).

Information is accumulated and generated at an enormous rate by powerful computers and data-acquisition systems. People complain that they are overloaded with information and are looking for better search engines and systems to help them. On the other hand, decision makers complain that they lack relevant information with which to make their decisions. This lack is largely due to the fact that much information is

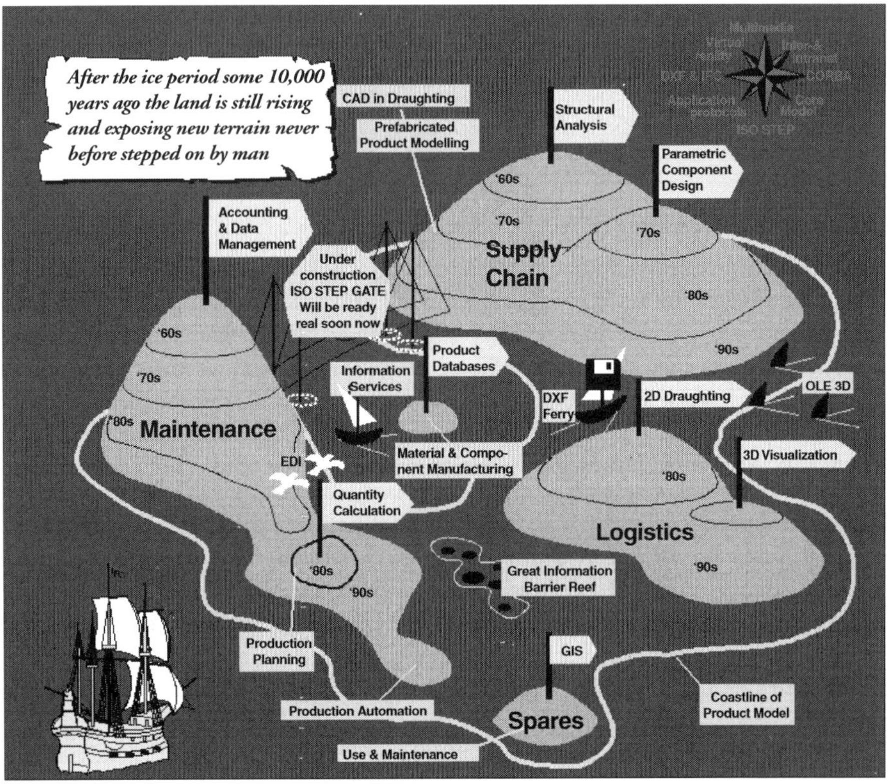

Figure 1.6-1 The islands of information

not shared in many parts of the manufacturing operation, even with business partners in the globally distributed environment. This isolation is often referred as 'islands of information'.

Islands of information can be visualised in many forms. For example, system diagnosis is related to diagnostics theories or implementations on some physical systems. The investigation of the characteristics of real time signals and their associated information has provided many links to the performance, behaviour and conditions of the machine from which the real time signals are captured.

The major task of support-system integration is to structure information in a way that can foster meaningful communication among distributed enterprises so that they can work more effectively. The principles of developing a support system information system are:

- Interoperability of information.
- Ensuring that the information that is created at one stage can be used without modification at the other stages (i.e. maintaining the validity of information).
- Incorporation of expert knowledge.
- Transformation of the knowledge that is normally associated with long experience and investigation to a mode that can be assessable by normal people.

- Customisation of service output.
- Creation of the ability to produce output that is specific to the customer so that special needs can be catered to without sacrifice by the customer. This ability should be obtained without undue cost burden.
- Improvement of quality of service.
- Use of new technology to ensure the output is controlled and hence the quality of the support service can be assured.

To ascertain a satisfactory level of integration, an organisation knowledge base must be constructed to selectively cull the data, analyse it and use it in a decision-support tool. The real value of an organisational knowledge base lies in the ability to develop actionable management information that attains results. Without such attainment, the data is not particularly useful.

1.7 Preparedness

A complex engineering system can have several utilisation states. In the working state, the system is operated by the user for delivering expected outcomes from the system. Reliability is the most important consideration in this state. In the idle state, the system is not operated for an outcome but should be ready for the next working state. The concept of preparedness drives the actions (e.g. maintenance planning, control) that are taken in the maintenance and logistics functions during this state of utilisation.

One of the key objectives of designing a support system is to achieve high reliability in assets that are mission-critical (e.g. weapons system, lifesaving equipment in an ambulance) and organisations that are measured by capability at any given time (e.g. defence, health care and emergency services, fire departments). Managing the link between reliability and capability as a part of the complete support solution requires a clear understanding of the capacity that the user anticipates at any time system is being operated.

The combination of readiness and sustainability is known as preparedness, which includes:

- *Readiness*: the ability of the complex engineering system to start and communicate to an operation within a specified time.
- *Sustainability*: the provision of all forms of support, including supplies that enable the complex engineering system to participate in an operation for a specified period that begins when the operation starts.

These concepts are linked diagrammatically in Figure 1.7-1.

Data and information flow within the support environment synchronises the maintenance team with the expected preparedness of the user. Typical questions from users include:

- Is the system ready to go at any given time?
- What is the standby capability at any given time (e.g. the number of combat-ready ships for deployment)?
- How much will it cost to maintain an agreed-upon level of preparedness?

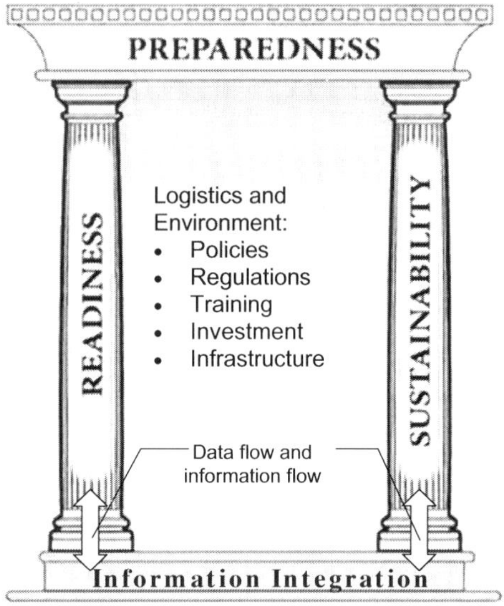

Figure 1.7-1 The pillars of preparedness

- Are all the supports in place to enable the system to fulfil its next mission?
- If some of the elements are not on board in the system, is the system prepared? What capability is lost without these elements?
- Is there a logistics process (e.g. an approval to release supplies) to enable the system's preparedness, both within the specified time and over the specified period?

As an example, when a ship is under 24 hours Notice for Sea, all crew members must be on board and be prepared to take the ship to sea within that period. Any of the ship's systems or equipment that has been shut down for defect rectification or planned maintenance must be repaired, reassembled and made ready within 24 hours. Information related to the questions listed previously would be critical to all parties involved in preparing the ship to put to sea on time.

Due to the widespread influences and complexity of interactions among the subsystems, not managing these interfaces well carries many risks. Managing the information flow at the interface between maintenance and logistics support to provide acceptable responses to the users is the key to customer satisfaction.

References

Dibia, I.K., Dhakal, H.N., Onuh, S., 2012. Lean: a continuous improvement philosophy in agile systems based on quality management principles. IJASM 5 (4), 370–389.

Higgins, A., Ferreira, L., Lake, M., 1999. Scheduling rail track maintenance to minimise overall delays. In: Proceedings of the 14th International Symposium on Transportation and Traffic Theory, 20–23 July, Jerusalem, Israel, pp. 779–796.

Johansson, P., Olhager, J., 2006. Linking product–process matrices for manufacturing and industrial service operations. Int. J. Prod. Econ. 104 (2), 615–624.

Karyotakis, K., 2011. On the optimisation of operation and maintenance strategies for offshore wind farms. PhD Thesis, University College London.

Metro Trains, 2013. Customer Service Charter, August, Accessible from Ministry of Defence (2005). Defence Industrial Strategy, Defence White Paper, December, 145 pages, United Kingdom.http://www.mod.uk/nr/rdonlyres/f530ed6c-f80c-4f24-8438-0b587cc4bf4d/0/def_industrial_strategy_wp_cm6697.pdf.

Ministry of Defence, 2005. Defence Industrial Strategy, Defence White Paper, December, 145 pages, United Kingdom. http://www.mod.uk/nr/rdonlyres/f530ed6c-f80c-4f24-8438-0b587cc4bf4d/0/def_industrial_strategy_wp_cm6697.pdf.

Mo, J.P.T., Zhou, M., Anticev, J., Nemes, L., Jones, M., Hall, W., 2006. A study on the logistics and performance of a real "virtual enterprise". Int. J. Bus. Perform. Manage. 8 (2–3), 152–169.

The life cycles of complex engineering systems

2

Chapter Outline

The philosophy of global business and supply chain is a key driver for aggressive manufacturers to take on the challenge of extending their activities to whole-of-life product-service systems. Inevitably, such companies will need to work together; meaning that they will need to contribute their resources to product design, manufacture, commissioning, support and disposal activities for businesses anywhere in the world.

Since the 1990s, industry and research organisations in Japan, Europe, the USA and Australia have been collaborating on research methods and tools that alleviate problems associated with the global business environment. The concept of virtual support enterprises was developed to describe the growing need for closer collaborations between companies and organisations. Although these entities in the virtual-support enterprise are not necessarily bound together by long-term formal arrangements, they are nonetheless required to produce a specific product or outcome together.

The outcome-driven virtual-enterprise environment naturally forces suppliers and customers to become part of the extended conglomerate of complex products. Adapting and adopting IT in ways that allow global virtual enterprises to achieve close working relationships among widely scattered companies and people provides tangible infrastructures for the effective operation of integrated virtual enterprises. The concept of life-cycle management is key to the success of such endeavours.

2.1 Complex engineering product life cycle

The concept of product life cycle comes from the traditional understanding of changes in the sales of a product at different periods, starting from a product's introductory phase. For complex engineering products, a more formalised and visible management

Engineering Systems Acquisition and Support

structure is required to manage the product life cycle. The challenge lies in consolidating and exploiting the knowledge in the virtual enterprise in order to fulfil the enterprise mission. The life cycle of a virtual enterprise begins when one or more of the partner companies sees business opportunities and invites the other companies in the business network to join the virtual enterprise. To support the operation of a virtual enterprise, the partner companies are not only required to provide resources and equipment for their part of the business but are also required to bring along knowledge and technologies that are appropriate to the virtual enterprise. Such knowledge and technologies are often classified as intangible assets because they are usually brought to the virtual enterprise by teams of people assigned to it, and because they will vary when different groups of people are involved. The highly unpredictable and invisible nature of these intangible assets makes it difficult for the virtual enterprise to create, control, maintain and share them among partners.

Complex engineering products are data-rich, knowledge-intensive outcomes. At first, the knowledge part is the intangible asset of the supplier. However, to effectively manage and operate the complex engineering product after commission, the owner or contractor doing the support work must acquire the same or an even better level of knowledge. To do this, it is necessary to understand the nature and characteristics of an intangible asset within the life cycle of the product with which it is associated.

The design of systems can vary greatly among vendors and concepts. In many cases, several consortia are involved and each brings its own expertise and expectations. The challenge to the system acquisition team is to establish suitable baselines for comparing different proposals. The acquisition process, which is already lengthy and complicated, is further complicated by the expectation that a complex engineering system must remain in service for years. This expectation is becoming more common.

From the owner's point of view, asset capability should increase over time to meet changing demands. To achieve this goal, typical support and services after acquisition of the asset should be included; two examples would be significant mid-life upgrades and continuous whole-system integrated logistics support. The philosophy of whole-of-life engineering development and support is to integrate design information with specific knowledge about implementation, manufacturing, servicing and decommissioning of the product. In addition, substantial planning is required to enable flexible delivery of service and logistical support to the asset from anywhere in the world.

2.2 Types of knowledge

In generic terms, knowledge is the internal state of an agent that has acquired and processed information that is based in previous experience (Nonaka, 1991). An agent can be a human being who is storing and processing information in his/her mind. An agent can also be abstract machine that includes devices to store and process information. There are two major components of the most widely accepted model of human knowledge: tacit and explicit (Zheng et al., 2000). Tacit knowledge consists of personal relationships, practical experience, shared values. Explicit knowledge consists of formal policies and procedures. (Some researchers regard formal policies and

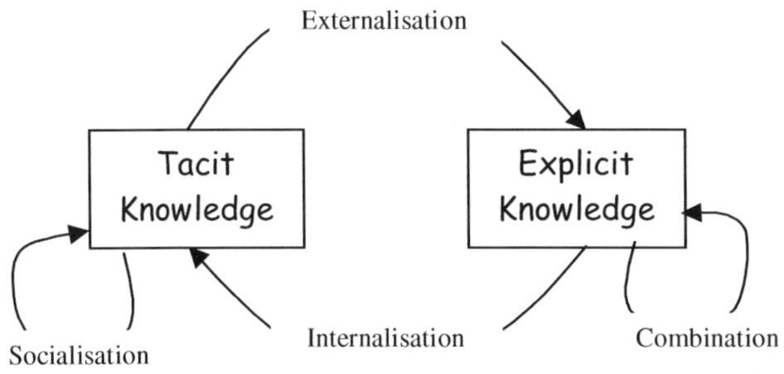

Figure 2.2-1 The tacit-explicit knowledge cycle

procedures as implicit, but this is a semantic issue.) Classification and capture activities are considered part of the externalisation process because they store the knowledge in a way that is explicit and accessible by other parties (Figure 2.2-1). After having created or acquired tacit knowledge, humans classify and capture their ideas by putting them on paper or in a computer.

Figure 2.2-1 presents explicit knowledge as something that can be expressed in a form that other people can acquire via access. Explicit knowledge is developed from the externalisation of concepts in documents, records, publications, procedure, instructions and so on. Information that has been captured in these formats is accessed and used by many parts of an enterprise, for various reasons. As this information is practiced, integrated, modified and combined, it undergoes changes; in turn, new knowledge is developed from these changes. This new knowledge may be explicitly expressed, recorded etc. or captured as memory. The parts captured in memory are internalised and have the potential to be shared among a number of people and organisational units. The term 'tacit knowledge' represents this type of enterprise wisdom, which is shared through a socialisation process (i.e. via contacts and interpersonal interactions). When tacit knowledge is externalised, the cycle continues as shown (Figure 2.2-1).

All forms of knowledge in an enterprise are intellectual assets of the enterprise. Stakeholders can use various tools and methods that allow the capture of tacit and implicit knowledge. The aim is to capitalise on the enterprise's intangible assets to create new knowledge appropriate to new projects. However, this process is difficult in a virtual-enterprise environment because each partner company may bring in a different set of tools and methods for knowledge externalisation. The virtual enterprise may have difficulty coping with these differences, which may themselves be incompatible.

It is clear that in order for any project to be successful, integration of numerous types of intangible assets among numerous organisations must be achieved. In a virtual enterprise, partner companies must contribute foreground knowledge to the virtual enterprise. Much of this knowledge can be explicitly defined and easily shared. It is more difficult to define and share background components that are implicit and tacit subjects that accompany the foreground knowledge unconsciously. Although individual companies may be able to capture and manage their own intangible assets in an *ad hoc* way, the implicit nature of tacit knowledge often results in real incompatibility. In such cases prevents, effective sharing of supporting knowledge components is prevented; this in turn poses serious obstacles to making all required knowledge available to all partners in the virtual enterprise.

2.3 Tools and methods requirements

Numerous interlinking activities that have to do with knowledge occur within the life cycle of a product. The concrete operational product is only one outcome of a huge amount of past knowledge. This knowledge is manifested in representations of the product on a variety of abstract levels, including conceptual/functional, design and manufacture. The knowledge elements within these levels are accumulated during various phases of the product life cycle (Figure 2.3-1).

The basic problem of the virtual enterprise is the creation of a knowledge repository that all partner organisations are able to understand, use and share. Due to the changing nature of the virtual enterprise in different parts of the virtual-enterprise life cycle, the term 'knowledge' will have different meanings. For example, during the after-sales support part of the life cycle, the virtual enterprise must understand the customer support process. This understanding must include the way this process should be carried out, the facts collected in the process and the inference activities that guide the customer through decision making at critical nodes in the process.

The following sections review the activities, information and tools at each stage of the product life cycle.

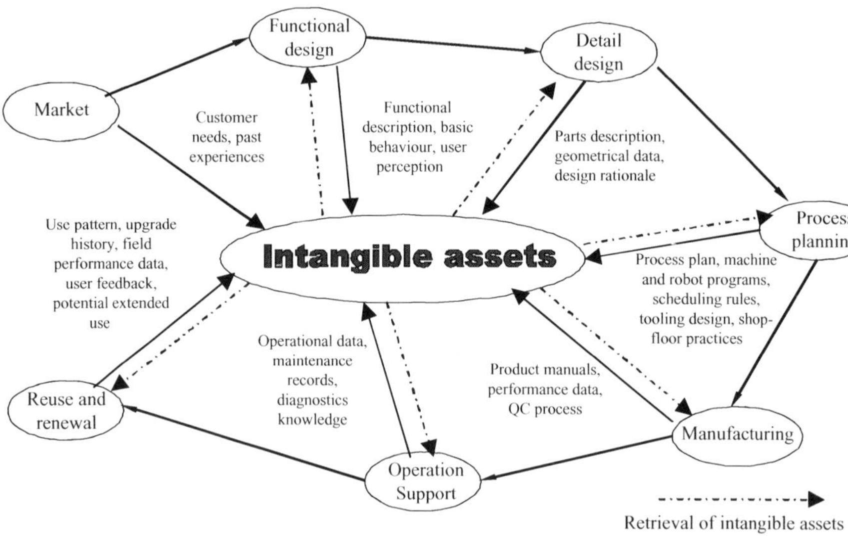

Figure 2.3-1 Accumulation of knowledge during different stages of product life cycle

2.3.1 Market phase

To compete in the global business environment, companies must collaborate across their company boundaries to form a virtual enterprise and bid for new projects together. This distributed bidding team involves members from the leading bidding company as well as its partners and possibly the client as well. During the bidding process, information that is necessary to the process must be captured in a specific structure that can be utilised as part of the corporate memory for supporting the preparation of bids quickly. In addition, the knowledge captured in the bid can also be used for contracting and total-product life cycle support. In a globalised engineering environment, a typical bidding process can be described in five stages (Figure 2.3.1-1). People from different parts of the world may be involved in the bidding. These may include bidding-team members and representatives of sales and finance divisions as well as customers and suppliers.

Figure 2.3.1-2 captures the process as well as the knowledge associated with the bidding process by showing an example of a bidding system that provides three layers of functionality. The lowest layer is a corporate knowledge-management platform that serves as the corporate memory. The middle layer is a team-working environment, such as a cross-corporate-boundary bidding-and-tendering project team, that supports the design and operation of a virtual enterprise. The top layer is the bidding workbench that enables the bidding team to prepare the cost estimation and technical proposal for the bidding-and-tendering document. Inevitably, the system requirement for such a global collaboration environment is based on the internet. At the core of the system, a database is required that comprises knowledge about products and production, knowledge of business partners and capability of risk assessment. Design of the database structure requires thorough understanding of the company's bidding process.

The corporate-knowledge management platform within a virtual-enterprise system attempts to manage both explicit and tacit knowledge. It enables the companies not only to know what they are already aware of, but also to be open to what they need to know. Thus, the platform enables the companies to plan for both procurement and updating of their knowledge.

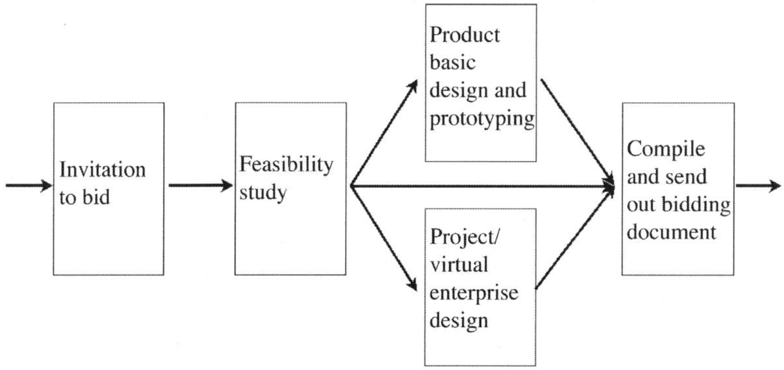

Figure 2.3.1-1 The bidding process

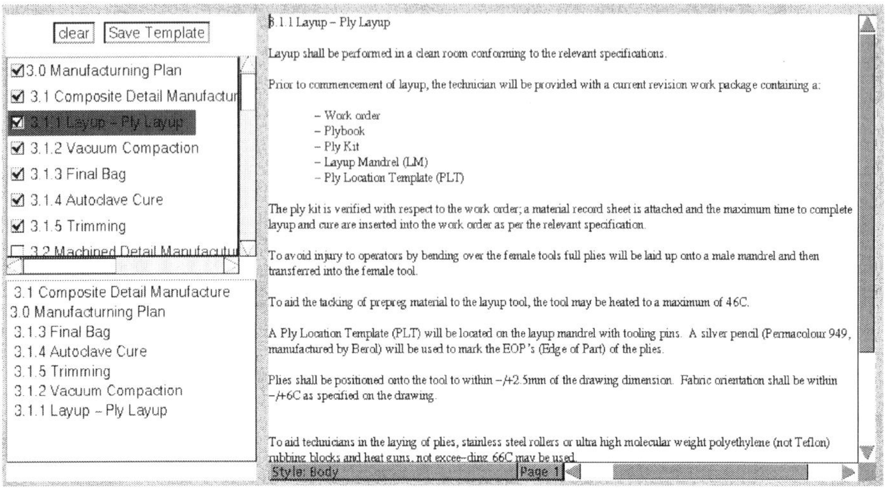

Figure 2.3.1-2 A sample screen of a bidding system

The bidding system provides a set of tools to prepare a bidding or tendering document for engineered products. It covers the whole bidding process including pre-bidding communication and decision support, bidding document configuration, online editing and reuse. This type of bidding document contains not only quotations, legal terms and conditions but also the statements of member companies' engineering capabilities, technical solutions, and quality-assurance procedures.

2.3.2 *Functional design phase*

After the market phase, if the bid is successful, the virtual enterprise moves into product development. Functional design is an essential operation that involves business partners who work closely together and consolidate customer needs (as identified in the market phase) into more-concrete product concepts (Kunz et al., 1996). The advantages gained from this teamwork are an increased commitment to the task, a larger body of shared experience and knowledge from which to draw and the generation of a greater number of problem-solving approaches and solutions. The creation of knowledge in the functional design phase is therefore largely through active communication and interaction among the participants.

The traditional way to foster communication among groups is by conducting meetings or issuing memoranda. Most manufacturing organisations still conduct meetings in much the same way they have done since the early 1900s. The traditional method of functional design can no longer support today's meetings requirements, however, because in a virtual enterprise, the business partners are geographically distributed. Under immense pressure to reduce time to market, companies form global teams from the best available resources of the virtual enterprise with the aim of bringing their products to market in the shortest possible time.

Table 2.3.2-1 shows different types of meetings that require different media to facilitate effective communication among the members of a virtual enterprise. For example, to carry out a session of detailed design analysis, all media channels must be used to ensure effective communication through the meeting. Apart from visualisation and audio tools for the design information, and document tools for displaying supporting information, a whiteboard tool is used to convey immediate ideas and innovative changes. The chat function is also ready to facilitate minutes taking and word-based data exchanges for which verbal channels are inefficient. By contrast, a progress meeting does not require detailed product information. Simply talking, supported by appropriate reports or documentation, is sufficient to achieve the objectives of this type of meeting. Likewise, a networking meeting does not normally present opportunities to explore ideas and exact data exchanges; therefore, such meetings require only some visual aids and audio-channel support.

The type of meeting convened depends largely on the status of the work at hand. Overprovision of communication resources does not necessarily serve the need of the circumstances. At the functional-design phase, the requirements of an effective support system are:

- Environment for personal interaction in real time; for example ability to enhance points of view clearly and concisely as well as formally and informally, as if participants are located in the same room.
- Friendly and responsive database management-and-searching capabilities that allow managers to access the same information, as if they are located at the same site.
- Fulfilment of the same set of operational functions globally as if the tools were local.
- Global operation, across international time zones and across company boundaries.

These requirements are now commonly found in computer-supported collaborative environments that many IT vendors can offer.

Table 2.3.2-1 Types of meetings and their media channel requirements

Need for communication/ type of meeting	Media channels used				
	Visual	Audio	Whiteboard	Document	Chat
Detailed analysis	X	X	X	X	X
Brainstorming	X	X	X		
Pre-design tender scope	X	X	X	X	X
Design review	X	X	X	X	
Problem solving-init.		X			
Progress		X		X	
Budget setting		X		X	
Budget review		X		X	
Strategy	X	X	X		
Networking	X	X			
Performance review	X	X		X	
Team selection	X	X		X	
Information transfer	X	X	X	X	

2.3.3 Detail design phase

During the detail design phase, designers must cooperate with many other functions in the global work team in all aspects of the design. The aim is to achieve simultaneous working of all functions within the virtual enterprise so as to shorten the time to market.

However, each partner company in a virtual enterprise has its own CAD system. Incompatibility among the different CAD systems used by different partners has imposed serious restrictions upon the design process. However, product-model data exchanges do not guarantee 100% data compatibility. When there is a need to combine different parts of the model, all component models must be located on one system. The process becomes difficult if the design must go through a series of trial-and-error tests in which the component models must be modified and recombined. The use of a standard for the exchange of product data (STEP) is an essential step to fostering the exchange of design information among CAD systems.

In addition, designers in different parts of the world still have the problem of communicating their design ideas to their colleagues elsewhere when those colleagues access the designers' CAD models. Information represented by CAD models is primarily explicit information about the product, which can be made accessible by remote CAD access facility via company intranet. However, the background information behind the design is often not included when the product model is distributed to the other members of the design team.

Attempts to capture design rationale have been made by some companies, but these approaches were limited to support for individual users only. For the virtual-enterprise environment, a design-concept information-management tool is required to capture design concepts on the basis of a bill of materials. Figure 2.3.3-1 shows a simple rationale-capture system that integrates with a CAD environment. When the CAD environment is integrated with STEP, the designer will have the tools for creating the design as well as entering vital information for the other partners in the virtual enterprise. This approach allows the sharing of design rationale among members of the global design team in this phase.

2.3.4 Process-planning phase

The intangible assets developed in the early phases of the product development cycle must be worked on. In the virtual enterprise, the processed data must be shared by individual manufacturing partners. Computer-aided process planning is a pivotal element in the integration of information from design to actual manufacturing.

It is common to use a feature-based CAD model-recognition system to process product characteristics in order to plan the manufacturing steps. A typical feature-based processing system (Figure 2.3.4-1) integrates the design stage through to the manufacturing activities; one example is a CNC machine that shows the examined functions and elements of the objects. Design information captured in a CAD data system is extracted according to geometrical characteristics and relationships. The extracted data is then processed in an object-oriented approach into a detailed description of the next manufacturing processes and machine programs.

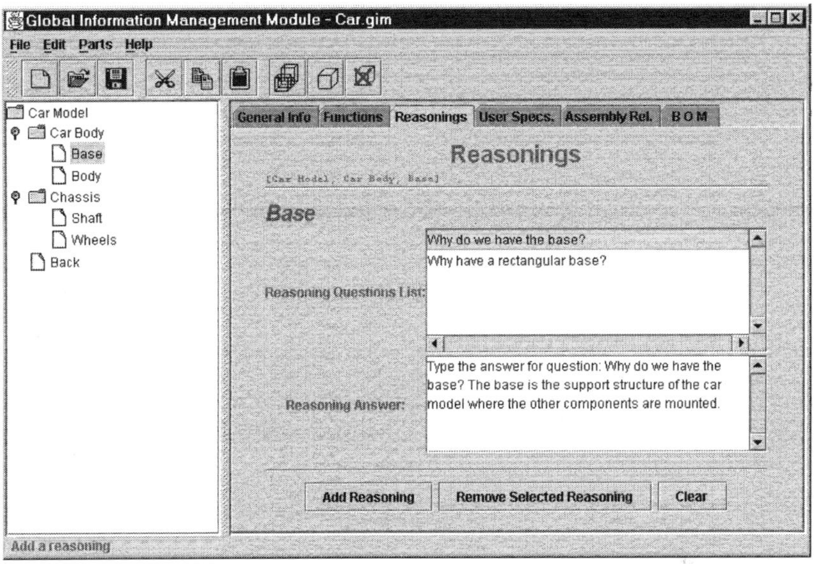

Figure 2.3.3-1 Reasoning capture

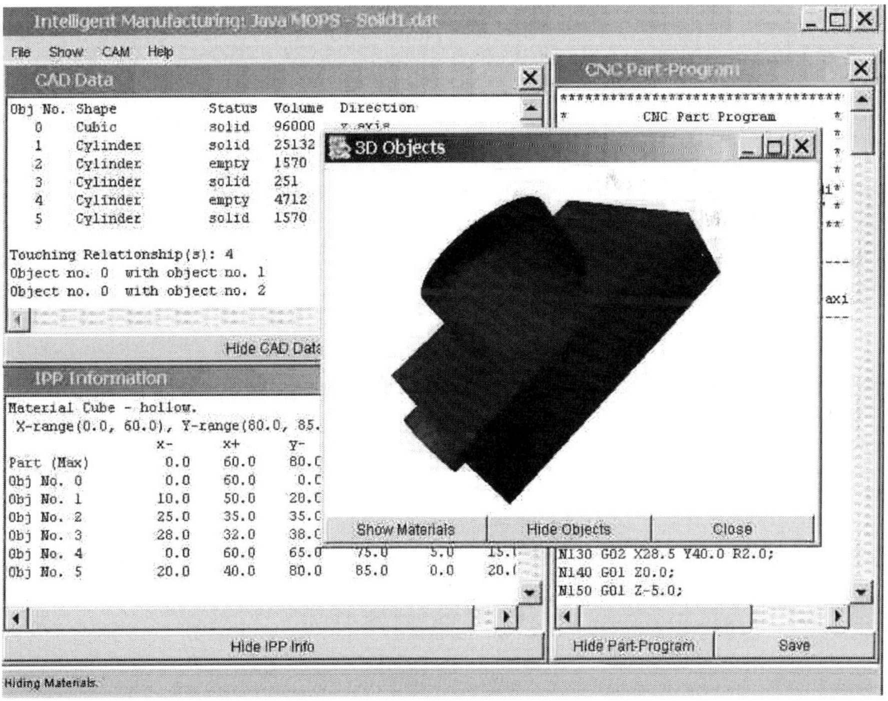

Figure 2.3.4-1 A simple feature-based planning system

In Figure 2.3.4-1, a simple component with cylindrical and rectangular bases is analysed to produce a CNC parts program that can be directly fed to a CNC machine.

2.3.5 Manufacturing phase

In the virtual-enterprise environment, the actual manufacturing is distributed to partners that have the most appropriate capabilities. Within the individual partner companies, scheduling and control are very much internal issues. Within the virtual enterprise, however, the successful integration of manufactured parts from various companies depends on quality management throughout the entire system. Within an individual company, the way to manage quality is by establishing a quality system (i.e. by establishing a systematic approach to plan, to do, to monitor and to control operations in achieving customer satisfaction). Thus, an individual company within the virtual-enterprise environment must make internal efforts towards quality achievement and also acquire quality materials, parts, and/or components from its suppliers. Therefore, establishing a quality system is crucial; inconsistent quality of incoming resources may cause difficulties in achieving quality products.

In quality management, the effective provision of documentation is vital to ensuring uniformity across all manufacturing units. First, system documentation demonstrates that the quality system is under control, clear and convincing. Second, the documentation on the quality records themselves is of authentic significance.

Typically, a company's quality system comprises a hierarchy of documents. This hierarchy forms a pyramid structure with quality policy at the top (Figure 2.3.5-1). Quality documents must be developed based on this top-level policy.

In a virtual enterprise, hierarchy of documentation is an intangible asset that must be shared and synchronised among all partners. Using an integrated process-modelling methodology, a web-based quality system that captures the quality system's manuals, procedures and work instructions can be developed. Depending on the system design, additional scripts to access a quality-control database can also be attached to the remote quality-management system. Such a knowledge repository is important not only for initial publication of the quality-system documents but also for modifications that

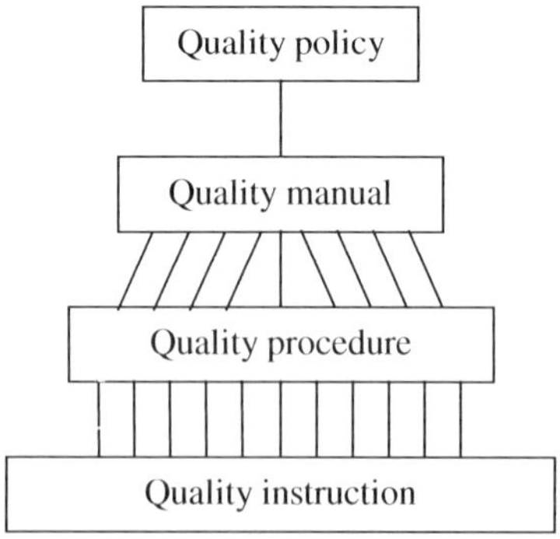

Figure 2.3.5-1 Hierarchy of quality system document

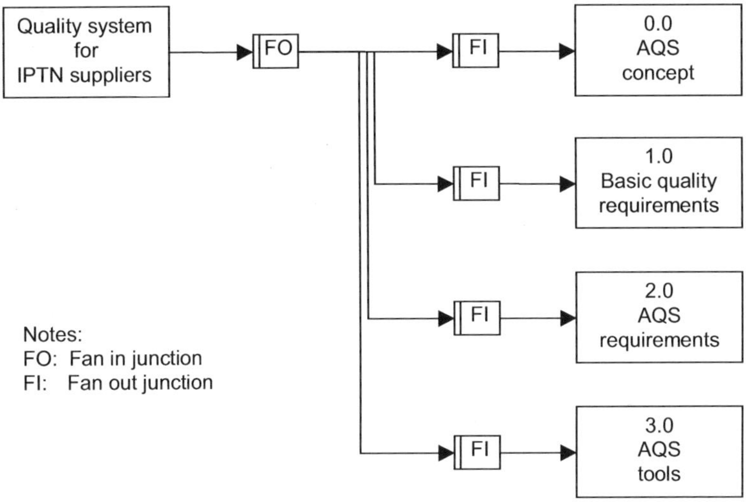

Figure 2.3.5-2 Main branch in quality-control IDEF3 process model

may apply during their life cycle. When change is required, the model is modified and the web-based support system is regenerated.

As an example, we model the quality-control function of an aircraft manufacturing company. Here, the process of quality control involves measuring and recording quality information about a product at specified stages of manufacture. Conformance of these quality measures against approved parameters is an indication of the achievement of-quality standard. The use of an Internet-based paperless documentation system is an important consideration in a virtual enterprise.

Figure 2.3.5-2 shows the main branch of the IDEF3 process model that has been created after the quality-control team captures the relevant steps in the manual quality-control system (the IDEF3 process-modelling techniques are explained in Chapter 7). The IDEF3 process system guides the users to select one of several different subsystems they wish to explore. A new user might decide to explore the Quality System Concept as well as the Basic Quality Requirements processes, whereas an expert user might choose to explore the Advance Quality System (AQS) Requirements process. Similarly, executives who will decide to implement the system might go straight to the summary in the Quality System Concept process, whereas their technical staffs or subordinates might be more likely to use AQS Tools in different scenarios as required.

The main objective of the Advance Quality System is to guide the user to explore the requirements sequentially, which they will do interactively with the system. After exploring the process(s) they choose, users will be linked to some possible pages that relate to those processes. There are two kinds of links: to the process that follows the current process, and to referred pages that contain related material and tools. With this arrangement, it is expected that users can be easily and systematically navigated through the system (Figure 2.3.5-3).

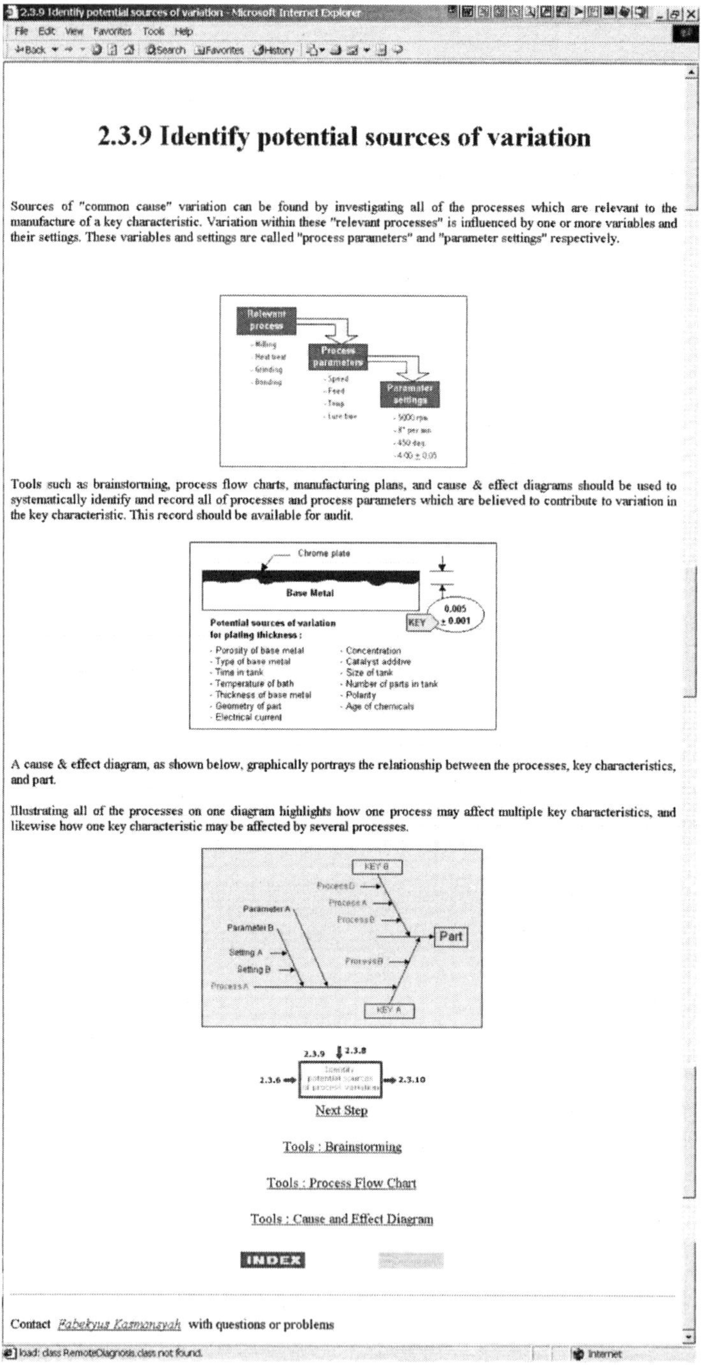

2.3.9 Identify potential sources of variation

Sources of "common cause" variation can be found by investigating all of the processes which are relevant to the manufacture of a key characteristic. Variation within these "relevant processes" is influenced by one or more variables and their settings. These variables and settings are called "process parameters" and "parameter settings" respectively.

Tools such as brainstorming, process flow charts, manufacturing plans, and cause & effect diagrams should be used to systematically identify and record all of processes and process parameters which are believed to contribute to variation in the key characteristic. This record should be available for audit.

A cause & effect diagram, as shown below, graphically portrays the relationship between the processes, key characteristics, and part.

Illustrating all of the processes on one diagram highlights how one process may affect multiple key characteristics, and likewise how one key characteristic may be affected by several processes.

Next Step

Tools : Brainstorming

Tools : Process Flow Chart

Tools : Cause and Effect Diagram

INDEX

Contact *Fabekyus Kasmansyah* with questions or problems

Figure 2.3.5-3 A typical page of the automatically generated quality system

2.3.6 Operation-support, -reuse and -renewal phases

Business activities related to a complex engineering product are extended beyond its sale and guarantee period. Therefore, companies try to harness post-sale business. For example, manufacturers wish to continue to advise and give guidance to their customers on how to operate their products, and when and how to carry out preventive maintenance. They also wish to help customers identify the reasons for eventual malfunctions and how to repair them if the equipment fails. These interactions with customers, which must be handled quickly and efficiently, require the development of several types of activities (Mo, 2003):

- *Documentation*. This is the major issue in supporting after-sales services on a global basis. The development of support systems for documentation should be focused on the automatic creation of remote-operation support systems from the company's intangible-asset database.
- *Training*. This is an important factor in successful after-sales services and support functions. For global after-sales services, it is necessary to consider training that must be conducted remotely.
- *Process tuning*. This is vital to ensure long-term problem-free operation for a complex engineering system. Remote-process monitoring aims at fast feedback from process experts to plant operators. The development of an expert-knowledge diagnostics-support system is necessary.
- *Built-in diagnostic procedures*. Many non-controllable factors are at play during system operation, such as changes in the conditions of the machine itself or the scenario in which the machine must operate. Not only are these factors unlikely to repeat exactly each time they occur, the system itself can break down in the most serious situations. Therefore, the system should contain built-in diagnostics procedures that will help the field technician as well as depot engineer locate the source of the problem.

The availability of high-bandwidth telecommunication networks on which large amounts of data can be transmitted electronically make it possible to support customers over long distances. Control-systems manufacturers have been successfully providing diagnostic services via networks to locate faulty boards and to put in replacement orders. Engine diagnostics and operational support for ships are being offered via satellite connections. Electronic distribution of documents for aircraft maintenance is offered by aerospace companies via broadband services.

These observations show that the remote-diagnostics system is built around a knowledge repository that consists of the field and engineering knowledge of the industry partner and its customers, as shown in the system architecture in Figure 2.3.6-1. In Figure 2.3.6-1, four user interfaces (Remote Machine Diagnostics, Remote Process Optimisation, Remote Knowledge Acquisition and Remote Operator Training) that interact with the user at remote location interrogate the Knowledge Repository to provide appropriate information to the user. In this way, all functions can easily be kept up to date.

The basic problem for developing the system is the creation of a knowledge repository. Knowledge in this context means the understanding of the whole customer-support process, including the way this process should be carried out, the facts collected in the process and the inference activities that guide the customer to make decisions at critical nodes of the process.

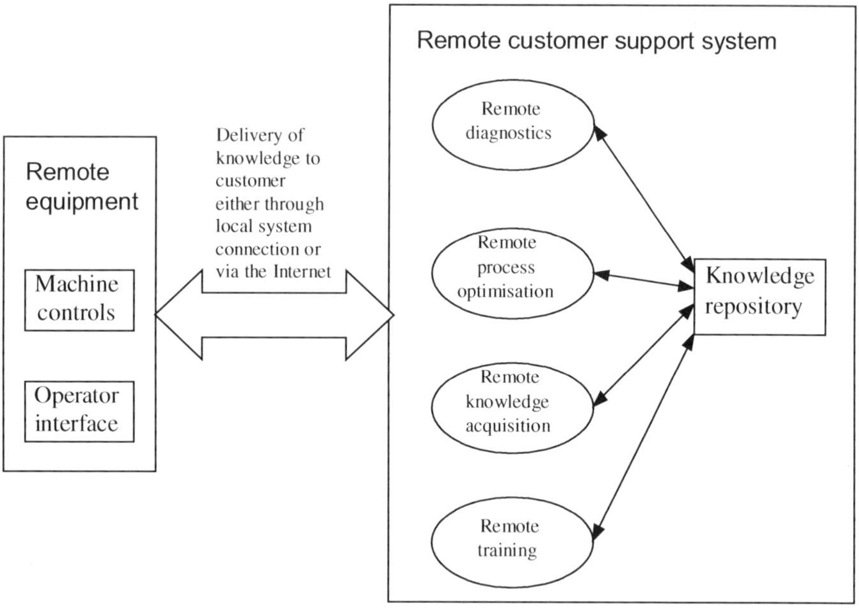

Figure 2.3.6-1 System architecture for Farley remote-operations support system

Visualisation of the process is just as critical to the support-service provider as to its customers because it is these people who must develop and maintain the system. In developing the system for such a complex process, the use of conventional expert-system languages has severe limitations because that technology depends on the involved people being thoroughly trained in artificial intelligence (AI) techniques. IDEF3's graphical representation, by contrast, can be easily understood by people. The customer-support process in that case can be described explicitly as a process map so that in-house experts in Farley can construct and maintain it themselves. Once this map has been created, it can be refined and continuously improved in collaboration with other experts and/or users.

Another example involves the explicit procedural characteristics of some of the remote customer-support process. The general process/dialogue with the customer in remote-access circumstances can be represented by a high-level IDEF3 model as shown in Figure 2.3.6-2.

The high-level model can be expanded as decompositions. The step 'Apply appropriate knowledge' includes suitable decomposition of the process into its components. Depending on the type of problem, the sophistication of the decomposition may vary. In data-communication troubleshooting, for example the steps of reasoning in the diagnostic procedure can be represented straightforwardly as an IDEF3 submodel. By contrast, the evaluation of cut quality is far more complicated and subtle because it involves the consideration of complex interactions between the machine parameters and the output-quality factors. The decomposition of this process therefore defines a link to an underlying rule-based expert system. One advantage of this approach is that

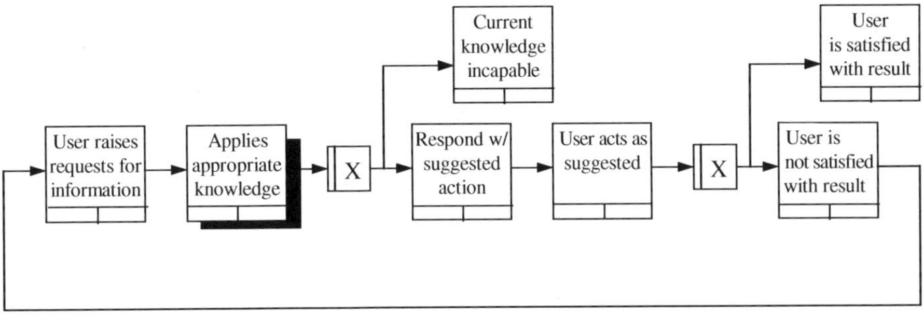

Figure 2.3.6-2 Remote-customer dialogue process

a significant part of the reasoning and fact-acquisition process can be represented explicitly by IDEF3; as a result, the complexity of the rule-based system can be reduced (given that the rules can be developed for a given fixed context in this manner, rather then in a context-independent fashion).

In this respect, the underlying rule-based system forms an essential component of the complete customer-support system. It is important to have a uniformity of approach across all the components of the total system (i.e. a process model for each component is required, regardless of the component's complexity). Once the knowledge is captured and properly modelled, realisation of the support system is handled by the software that generates the Internet-based graphical user interface.

2.4 Whole-of-life engineering

No two systems-acquisition-and-support projects are the same because each system must meet different user requirements (Kucza & Gebauer, 2011). Due to the highly individualised nature of the acquisition and support of complex engineering systems, it is an extremely knowledge-intensive, labour-rich business that requires the development of new systems in the form of design certification, digital manufacturing, integrated testing, customer support centres, information infrastructure, human organisations, multidisciplinary engineering teams, supply chains and distribution networks. Operations of these engineering units in a globally distributed environment present significant challenges to engineering-system developers.

Configuration management is the management of a system's elements in a particular arrangement to deliver its designed capability. The configuration is more than just an asset register; it contains information that relates to the maintenance, movement and problems experienced with the configuration items. Generally, a configuration management plan is developed to ensure successful implementation of a configuration. It provides details of how the system-support team will manage the control of configuration items on the in-service assets and the changes that may be commissioned during scheduled maintenance. The plan defines the policies and procedures for configuration management and the infrastructure necessary to implement them.

Changes in the configuration of an in-service system are usually implemented by the maintenance function. Control of the configuration, in the form of an as-maintained baseline of parts used, is essential to good parts management. Because an as-maintained baseline differs from the corresponding as-built baseline, the interface between maintenance and configuration management consists of the flow of information about the processes, supplies and competency requirements for configuring the fielded system.

Maintenance activities are informed by the system-configuration requirements that are needed to support the next mission. Inadequate configuration control of the information will result in a situation where the overview or knowledge of the actual parts used as replacements as well as the processes to change the configuration and the skills required for quality assurance are unknown. For this reason, changes to the as-maintained baseline can only be made using an approved process of authorising changes to parts and equipment. The keys to good parts management are accurately knowing the configuration and managing it accordingly.

Furthermore, configuration management and change control must be linked because this linkage is essential for capturing the actual changes that have been made to the configuration of the engineering system. Such changes may also include modifications that have been made due to operational needs. Maintenance data and information is needed to ensure full and exact knowledge of the configuration of an engineering system at any given time in its service life.

The configuration management process covers

- Determination of the boundary between what is contractor's responsibility and how is it managed by the contractor and what requires regulatory authority oversight and involvement.
- Defect reporting that covers all aspects of defects, ranging from published data in CM system to asset issues.
- Documentation updates that cover all technical publications and maintenance.
- Concession and permits (deviation and waivers) against any configuration baseline.
- Permanent configuration changes to any configured baseline.

To change a configuration, support engineers must generate engineering change proposals (ECPs) that pass the completeness test and consider the whole-system life cycle. Impact analysis of a proposed change requires consideration of multiple phases of the system life cycle. Thus an ECP must consider:

- Different system instances at different build states (e.g. all ships of the same type are not always identical) for significant periods of time.
- Need for upgrade or disposal of spare-part inventories.
- Staff training for upgraded systems.
- Other factors suggested by the elements of capability.

When developing an ECP, beware of imposing existing current constraints on a new capability. Acquirers must make a judgement that relates to imposing existing constraints on the development of a new system versus investing to remove those constraints in order for the new system to best meet operational needs. For example, a change proposal for a defence system imposed constraints by mandating the use of in-service communications equipment within the system. However, it was prohibitively difficult and error-prone for operators to set up and maintain complex communications networks for the in-service system in light of the fact that the communication interface would eventually be replaced. It probably would have been more cost effective to not mandate the in-service communications system.

Acquirers will often apply their current practices, experiences and procedures to new capabilities. For example, acquirers would be able to use some prior knowledge (but not all) for a large naval vessel that has massive transport capability and thus should be viewed as a transport ship instead of a fighting platform.

Validation of the support requirements used in the acquisition phase (design for sustainability) is an essential part of an in-service ECP. As the system transitions into service (and as part of monitoring ongoing supportability), the support team should measure support capability (i.e. they should verify the support system). For example, an airline introducing a new aircraft into service should ensure it has some means to perform essential corrective maintenance along the aircraft's intended routes. Before the system is transitioned into service, it is important to objectively measure the support elements of capability, which include trained staff, spare-parts availability, rates of consumption etc.

For operational requirements, it is important to ask how the users will actually use the system. Especially in the case of an enduring system with a very long life, users will use the system as they want to, not as the designer intended. It is also important to ask whether the use pattern of the system is known. For example, ANZAC ships are designed for the Southern Ocean environment but have turned out to operate extensively in the Middle East. This has caused support challenges in terms of distance from depot, environmental issues etc.

Other relevant questions to ask are:

- How frequently will the system be used?
- Under what circumstances will the system be used?

As the world is becoming more environmentally conscious, consideration for the disposal of hazardous components and waste becomes necessary. The following are factors that should be considered:

- Ecological requirements.
- Social requirements.
- Economic feasibility.
- Skill sets of available users.
- Availability of operation and maintenance personnel.
- Administrative delays.
- Storage-capacity constraints.
- ISS facilities limitations.
- Need for capability-growth margins.
- Expected operational life of system.
- Utilisation requirements.
- Maintenance down time.
- Failure rates.
- Maintenance levels.

The optimised solution design should have adequate requirements elicitation (requirements collected from multiple view points), complete life cycle costing modelling, alternative solutions identification and evaluation and an impact assessment on existing capabilities and infrastructure.

Once an equipment type is in service (indeed, even when it is in manufacture), any subsequent engineering change is heavily constrained by the original system design and by the in-service environment. Because it is usually impossible to change the fundamental architecture of the system, changes must be consistent with that architecture. However, the asset does not exist in isolation: there are always other elements of the system architecture that contribute to the capability delivered by the integrated system. Elements that contribute thusly can be abbreviated as TEPIDOIL:

- *T*raining
- *E*quipment
- *P*ersonnel
- *I*nformation
- *D*octrine and concepts
- *O*rganisation
- *I*nfrastructure
- *L*ogistics

These elements should ideally be harmonised because they are codependent in various ways. For engineers, it is tempting to address only the equipment element of the capability, but doing so can be highly detrimental to the overall system capability. In fact, taking the systems view of capability opens up other non-equipment options; for example, changes to training on equipment use may obviate the need to make modifications to the equipment. The engineer should be fully aware of all these elements of capability in order to achieve an optimum business solution.

These capability elements are often known as fundamental inputs to capability. They must be integrated and managed holistically within a defined or constraining architecture to realise and sustain a deliverable capability. A system designed to only meet operational needs cannot be a valid system design if it is not supportable.

References

Kucza, G., Gebauer, H., 2011. Global approaches to the service business in manufacturing companies. J. Bus. Ind. Market. 26 (7), 472–483.

Kunz, J.C., Luiten, G.T., Fischer, M.A., Jin, Y., Levitt, R.E., 1996. CE4: concurrent engineering of product, process, facility, and organization. Concurrent Eng. Res. A. 4 (2), 187–198.

Mo, J.P.T., 2003. Case study – farley remote operations support system, Chapter 21. In: Bernus, P., Nemes, L., Schmidt, G (Eds.), Enterprise Integration Handbook. Springer-Verlag, 2003, pp. 739–756, ISBN 3-540-00343-6.

Nonaka, I., 1991. The Knowledge creating company. Harvard Bus. Rev. (November–December), 96–104.

Zheng, J., Zhou, M., Mo, J.P.T., Tharumarajah, A., 2000. Background and foreground knowledge in knowledge management, Global Engineering, Manufacturing and Enterprise Networks, Proceedings of fourth International Conference on Design of Information Infrastructure Systems for Manufacturing, November 15–17, Melbourne, Australia, pp. 332–339.

Systems acquisition principles

3

Chapter Outline

3.1 Systems-engineering approach

Adoption of the system-engineering approach can give stakeholders a clear picture of what the real-life industry project is and how new system design is developed to meet user expectations and requirements. Many influential factors such as climate change, oil-price uncertainties and population growth may come into play. The system-engineering approach specifies many stages of the system-design process (Figure 3.1-1). As each stage progresses into the next stage, a documentation and approval process is required. This process, however, is not a one-way path. As shown in Figure 3.1-1, the design concept and detail plans (at the left-hand side of the V process) are validated by testing, verification and evaluations at the later stages (at the right-hand side of the V process). During acquisition phase, we focus on the left-hand side.

An acquisition policy should have five key characteristics: flexibility, responsiveness, innovation, discipline and management that is both streamlined and effective.

- *Flexibility* refers to the need to structure each acquisition program according to the set of strategies, documentation, reviews and phases that make sense for that program. According to the nature of the items being acquired, acquisition programs will be structured in different ways.

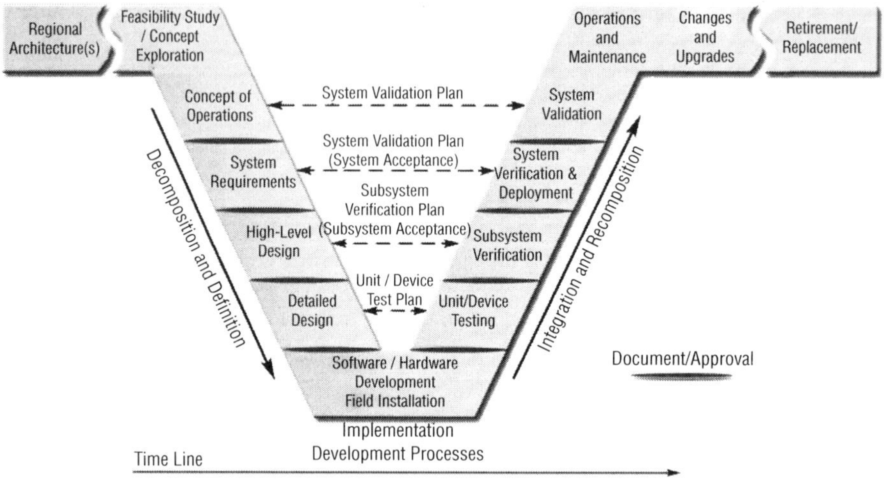

Figure 3.1-1 Systems-engineering process (International Council on Systems Engineering, 2011)

- *Responsiveness* refers to the need to react quickly to new and advanced technologies. Phasing of capability needs and evolutionary acquisition strategies designed to leverage future technologies should be timed as they become available, to help increase responsiveness.
- *Innovation* is required to constantly improve and streamline acquisition practices. In particular, electronic solutions and best commercial practices are encouraged wherever possible. Overall, the goal is to reduce cycle time and cost while improving teamwork.
- *Discipline* means ensuring compliance with regulations; establishing goals for cost, schedule, and performance for the life cycle of the program; and effectively managing deviations from those goals as well as baseline parameters and exit criteria.
- *Streamlined* refers to the synchronisation of steps in the acquisition process to remove any potential delay.

3.2 User requirements

Understanding user requirements is an integral part of systems design and is critical to the success of a support system (Shen et al., 2004). The ISO 13407 standard specifies needs and requirements. A customer-friendly support-system design begins with a thorough understanding of the needs and requirements of the customer. The benefits can include increased productivity, enhanced quality of work, reductions in support and training costs and improved user satisfaction. Requirements analysis is not a simple process. Particular problems faced by the system designer include:

- Addressing complex organisational situations with many stakeholders.
- Participants thinking along traditional lines that reflect the current system and processes, rather than being innovative.
- Customers not knowing in advance what they want from the future system.
- Rapid development cycles, which reduce the time available for user-needs analysis.

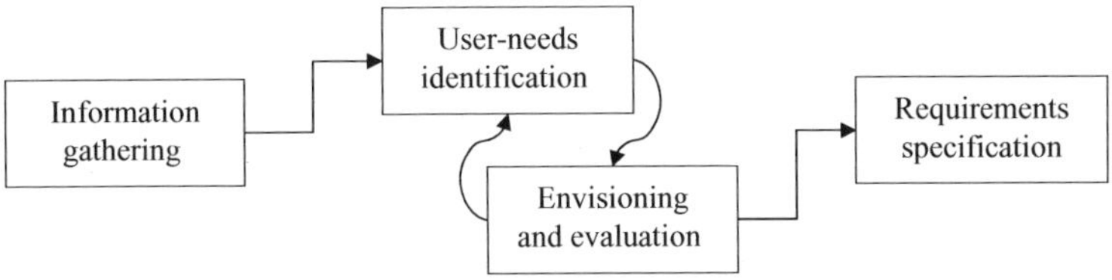

Figure 3.2-1 Elements of user requirements

A typical requirements capturing process encompasses four elements (Figure 3.2-1).

Requirements should have certain characteristics in order to be effective. Writing the requirements well is as important as determining the right ones. Requirements should be:

- *Solution independent*. This means that requirements should not specify a solution to the problem; they should specify what needs to be done, but not how it will be done.
- *Complete*. Requirements must cover all areas of concern, including all phases of the product life cycle.
- *Clear*. Requirements should not leave anyone guessing what is required.
- *Concise*. Unnecessary requirements should be omitted, and the wording of requirements should be concise. Do not bury the requirement in unneeded text.
- *Testable*. When possible, quantitative (numerical) limits, tolerances, ranges and intended values should be indicated. Testable requirements can be measured in order to determine if the design goal is met.
- *Traceable*. It should be possible to trace a requirement back to the rationale from which it was derived and forward to its implementation in the design.

3.2.1 Information gathering

The first step in user requirements analysis is to gather background information about the system to be supported and how the system is used in service. The following data sources can be explored:

- All the users and stakeholders who may influence or be impacted by the system. User groups may include end users, supervisors, installers and maintainers. This information can be gathered by stakeholders.
- Published sources such as research reports, census data and demographic information that throw light upon the range of possible user markets. Market intelligence can find these sources.
- Context-of-use analysis, which basically means information about the characteristics of the users, their tasks and operating environments. Usability, accessibility and social-acceptability factors may also be considered.
- Task analysis, which shows what a user will do to complete a task. The common tool is a flow chart of the sequence of activities and the associated inputs and outputs.
- Rich pictures consisting of simple figures that show peoples' roles, power structures, communications and reporting mechanisms.
- Field study involving an investigator unobtrusively observing or making a video recording as the user works.
- Diaries kept by the user as a record of activities and events over a period of time.

3.2.2 User-needs identification

In addition to collecting user data, user needs must be identified. Defining these needs is largely influenced by user opinion. A number of methods exist for identifying them.

- *User surveys*, which means written questions sent to a sample population of users. Surveys are normally composed of a mix of closed questions (i.e. with fixed responses) and open questions (i.e. the respondents are free to answer as they wish).
- *Focus groups* that bring together a cross-section of stakeholders in a discussion-group format.
- *Interviews*. In this commonly used technique, users, stakeholders and domain experts are questioned to gain information about their needs or requirements in relation to the new system.
- *Use cases*, which means symbolic models that describe examples of future use of the system and serve as aids to understanding and clarifying user requirements in specified circumstances.
- *Future workshops*. These are focused sessions in which the participants are requested to express their goals for the future.
- *Evaluations* of an existing or competing system to provide potential user requirements.

3.2.3 The use-case model

Use-case modelling is a common method in modern software development to describe a system's software requirements. However, the process is generic to model interactions between human and systems, or between systems. From a use-case model, it is possible to determine not only other representations of the system at the same level of abstraction but also representations at lower levels of abstraction. A use-case model presents an external view of the system that is also referred to as a black-box view. A use-case model consists of use-case diagrams depicted in UML, and use-case descriptions (Eynard et al., 2004). The UML model depicts the use case, actors, communication associations between actors and use cases and use-case relationships (particularly the extends and includes relationships). The UML notation does not address the use-case description, which is in many ways is the most important part of the use-case model (Figure 3.2.3-1).

The use-case model is the starting point for the creation of test-item specifications. With the use-case modelling approach, the functional requirements of the system are described in terms of actors and use cases. A use case contains a main scenario and (usually) several alternative scenarios. A use-case main scenario describes a sequence of interactions between one or more actors and the system. The alternative scenarios describe a sequence of inputs and actions that differ from a typical use of the system, such as for error handling. Each use-case scenario can in turn be described with an interaction diagram that shows the sequence of interactions between the use-case actors and the system.

3.2.4 Envisioning and evaluation

Once an initial set of user requirements has been developed, it is important to develop a prototype to illustrate them. User feedback can then be obtained on the prototype

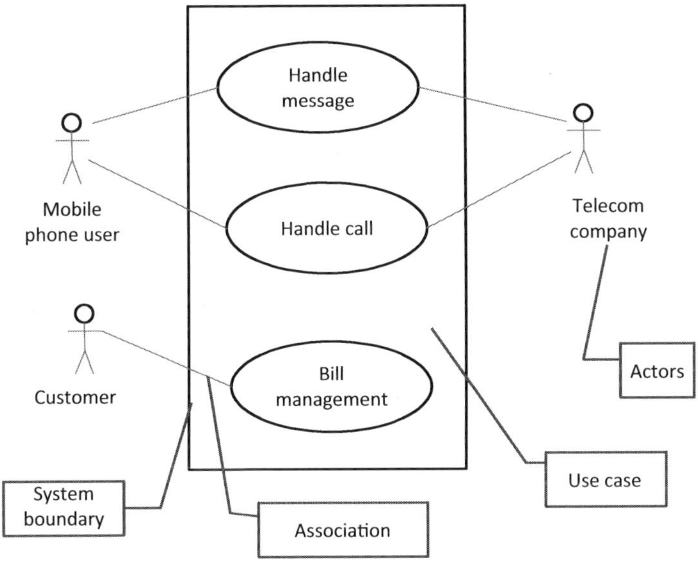

Figure 3.2.3-1 Use-case model

to validate and refine the user requirements. A number of useful techniques can be employed in this process, including:

- *Brainstorming sessions* that bring together design and task experts to inspire each other in the creative, idea-generation phase of the problem-solving process. These sessions are used to generate new ideas by freeing attendees' minds to accept any idea that is suggested, thus allowing freedom for creativity. This technique has been widely used the early phases of design. The results of a brainstorming session are, it is hoped, a set of good ideas and a general feel for the solution area to meet user needs.
- *Card sorting* is a technique for uncovering the hierarchical structure in a set of concepts by asking users to group items that are written on individual cards. For working out the organisation of a website, for example users would be given cards with the names of the website's intended pages and asked to group the cards into related categories. After gathering the groupings from several users, designers can typically spot clear structures across many users. Statistical analysis can uncover the best groupings from the data when it is not clear by inspection.
- *Affinity diagramming* is a related technique that can be used for organising the structure of a new system. It allows participants to work as a group. To uncover the structure and relationships in a domain, designers or users write down items such as potential screens or functions on sticky notes and then organise the notes by grouping them. Affinity diagrams are often a good next step after a brainstorming session.
- *Storyboards*, also called presentation scenarios, are sequences of images that show the relationship between user actions (i.e. inputs) and system outputs. A typical storyboard will contain a number of images of features (e.g. menus, dialogue boxes and windows).
- *Prototyping* is a pseudo-model that users can plan within a hypothetical user environment. The difficulties encountered by the user, along with user comments, are recorded by an observer.

- *Allocation of function*, which is an important element for many systems. The aim here is to specify a clear system boundary by establishing a range of options to identify the optimal division of labour, in order to provide job satisfaction and efficient operation of the whole work process. User cost-benefit analysis can then be carried out to determine how acceptable each user group will find the new arrangement. Task allocation charts and cost-benefit analysis are more useful for systems that affect whole work processes than for single-user, single-task products. They also provide opportunities to rethink system design or user roles so as to provide solutions that are acceptable to all groups.
- *Design guidelines and standards*, which are referred to by designers for guidance on ergonomic issues associated with the system being developed. Style guides embody good practice in interface design. Following a style guide will increase the consistency between screens and can reduce the development time.
- *Parallel design sessions* involve a few small groups of designers working independently to generate a range of diverse solutions. The aim is to develop and evaluate different system designs before choosing a solution (possibly drawing from several solutions) as a basis for the implemented system.

3.3 Requirements analysis

Requirements analysis determines if and how the requirements can be met. Analysis data should be:

- Expressed as solutions to scenarios or customer problems, or as benefits.
- Expressed in customers' language, derived from customers' information.
- Clear to anyone in the development organisation.
- Expressed as positive statements whenever possible.
- Qualitative terms, not numbers (customers buy benefits, not features).
- Based on definitions that anyone will understand.

User requirements is about capability and the customer/user. What the system must do to deliver capability is a common inquiry that is usually answered in this manner: 'As a result of acquiring this system, the user will have the capability to do X'. The system requirements can be described in a similar manner: 'In order to give the user X capability, the system will do A, B, C'.

Good requirements are:

- *Complete.* All information for implementation, individually and as a set, is provided.
- *Unambiguous* and *clear.*
- *Consistent* and not in conflict with other requirements.
- *Demonstrable* and *verifiable.*
- *Expressed as 'what'*, not '*how*'.
- *Correct.* They contain no error that affects design.
- *Written in ways that clarify customer/user needs, expectations and constraints.*

In designing a system, there may be many types of constraints. The system that is to be supported may comprise many black-box elements. Elements of a system are managed in terms of form, fit and function, as well as their compatibility to other parts of the system. Sometimes, however, very little is known about the internals of those elements (hence the term 'black box'). The owner will typically apply risk-management principles to determine how best to support the product. For example:

- An aircraft MRO business will manage the maintenance of the aircraft engines differently than the galley stoves.
- A mine may outsource maintenance of its trucks completely through a service-level-agreement.

The people who are operating the system usually do not know as much as the people who built it. Often important knowledge has been lost, is dissipated or is otherwise not available to support the system. Engineering disasters can occur because lack of knowledge carries safety implications, such as,

- Maintenance procedures are not strictly followed.
- Unauthorised or poorly engineered modifications.
- Changes in the environment that are not well understood until after negative consequences are realised.

A system's operational concept should identify the operating environment and support environment constraints. These will impose constraints that the system solution will need to address. Impact on existing support systems (staff, facilities etc.) should be analysed.

The support constraints will put limits on the allocation of maintenance tasks to the different maintenance levels (e.g. organic, intermediate, depot, OEM) and other factors (e.g. infrastructure and skills). For example:

- Due to remoteness, a mine-system solution will include a local fuel depot for trucks and other equipment.
- The introduction of a new aircraft type to an airline will require type-training of existing maintenance staff and possibly changes to facilities.

Another consideration related to the reliability and usability of the system is its emergent in-service characteristics. Different system instances may be at different build states. System instances are committed to operations, which places constraints upon their availability for upgrades or maintenance. One instance of a system (e.g. a mine or a ship) may be retired as another instance is being transitioned into service. Over time, a system's operating environment changes (e.g. a second, different system is brought or installed in close functional proximity to the first system) and potential interactions between the two need to be understood and managed (e.g. a township develops around a mine site). These constraints must be taken into consideration when applying systems-engineering principles to the support solution.

Similarly, changes to the support environment (e.g. external legislative requirements) must be taken into account (e.g. OHS, environmental changes, litigation trends, government and political policies). Changing legal obligations (e.g. duty of care, OH&S, equal opportunity) may result in having to change elements of a system and its support system. Generally, the community has a growing aversion to safety risks. Thus, one example of such a change is asbestos heat-shielding, a piece of support-environment equipment that is no longer permitted and must therefore be replaced. But another example is equal-opportunity clauses, which may mean that maintenance tasks that require superior upper-body strength can no longer be specified for or restricted to men only.

Cultural/generational factors such as end-user drifts or changes in behaviour, attitude, competency, work culture etc. may also come into play. People who were born in the final decades of the twentieth century have different values and expectations than previous generations; today's youth are IT-savvy and have grown up being able to communicate anywhere and anytime. Because operations in remote places or on a warship may come as a cultural shock to such people, owners of mines and warships may need to make provisions for new, younger generations of staff and workers.

3.4 System specification

System specification is a requirements specification that clarifies

- Formal, contractual documents.
- Baselines of development and document requirements.
- Basis of system acceptance.
- Military standards that define several types of specifications.

The purposes of system specification are to

- Serve as a contract between the customer and the developer.
- Serve as a source of test plans.
- Specify project goals, plan-development cycles and increments.

Specifications that focus on performance state requirements in terms of required results and provide criteria for verifying compliance. They do not state unnecessary methods for achieving results. This compares to traditional (detailed) specifications that state specific design requirements (including materials) are to be used and how a requirement is to be achieved (e.g. the methods for its production).

The flow from user requirements to system requirements should be clear and documented. The move to system requirements entails a move from the language of the customer to the language of the engineer. There will be some overlap; for example some user requirements may be more technical in nature and may even constrain the design.

Regardless of the type of requirement being described, the prevailing fashion is towards performance-based specifications in the general sense. They specify *what* the system must do instead of prescribing *how*.

Other factors to consider are:

- Performance specified at the subsystem or component level.
- Requirements that are not measurable or verifiable.
- Statements that constrain the solution to a single solution (e.g. 'fabricated from xyz').
- Requirements that are vague.
- Orphan requirements.
- Specifications that are solely to define performance and are not directly tied to scope/mission.
- The use of mandatory requirements that are not justifiable.

3.5 Tender management

Tender evaluation is one of the most important activities in the procurement process (Noor et al., 2008). Proper tender evaluation ensures value for money for the buyer and that the asset will demonstrate compliance with accountability and probity requirements. A standardised approach to the conduct of procurement (including tender evaluation) will reduce errors as well as ensure the maintenance of appropriate standards of probity and the achievement of improved outcomes.

Procurement for complex engineering systems can be classified into three broad classifications: simple, complex and strategic. In the category of simple procurement, a risk assessment commensurate with the size and complexity of the procurement has declared the overall level of risk and complexity to be low. The risks involved in procurement increase as the value and complexity increase.

For all three procurement categories, tender evaluation represents a critical stage of the broader procurement life cycle – one that is aimed at achieving value for money. The procurement life cycle encompasses planning, getting into contract and contracting management through to disposal. Tender evaluation can only be successful where earlier activities have been conducted appropriately (e.g. development of the acquisition strategy/procurement method, release documentation, tender-evaluation plan). Tender evaluation can therefore be successful only where it contributes to value for money holistically across the entire procurement life cycle.

3.5.1 *Principles of acquisition tenders*

All procurements must be conducted in an efficient, effective, economical and ethical manner to ensure that the public has the utmost confidence in procurement processes. The main mechanism in the acquisition process is the use of tender.

The following key principles must govern the conduct of tender.

Fairness	No tenderer must be given an unfair advantage or be disadvantaged. This means that all tenderers are given the same information and afforded an equal opportunity to succeed.
Confidentiality	Appropriate measures must be taken to protect the confidentiality of tenders and information related to tender evaluation. Internal distribution of information should be on a strictly need-to-know basis. Tenders should be treated as commercial-in-confidence.
Tenderer communications	All formal and informal communications with tenderers throughout the procurement process should be channelled through a single point of contact. A written record should be kept of all oral and written communications.
Tenderer clarification	Clarification should be sought where information contained in a tender is unclear or open to differing interpretation and is potentially relevant to the final decision. Clarification can be sought orally (e.g. by phone) or in writing (e.g. by email); in any case, an accurate written record must be kept of all questions and their answers. When tenderers respond to clarification questions, they must not be allowed to revise or enhance their original offer.

Probity	All officers involved with tender evaluations must exercise the highest standards of probity and fair dealing, including by ensuring an absence of bias or favouritism throughout the process and by promptly declaring and managing any conflicts of interest.
Documentation	A permanent and easy-to-follow audit trail must be maintained for all procurements. All key steps and decisions must be promptly and accurately documented in a logical sequence, using clear and concise language to ensure that the process can withstand challenge and external scrutiny.

3.5.2 Tender evaluation

Tender evaluation must be undertaken by staff with appropriate skills, training and knowledge that is applicable to the nature of the procurement.

For simple procurements it is generally sufficient and appropriate for a single staff member to undertake the evaluation. However, for procurements farther along the complexity and risk spectrum, it may be appropriate for a panel of two to four members to conduct the evaluation. A recognised advantage of using a panel is that it takes away some of the bias risk that may accompany tender evaluations.

When a tender evaluation panel is used, it should be constituted and members advised of their respective roles and responsibilities before tenders are sought. A single member should be chosen as chair; this person should have delegation responsibility for endorsing the final source-selection recommendation. The roles and responsibilities of each panel member should be agreed and documented before tenders are sought.

Where a tender evaluation panel is used, it is best practice for each team member to assess each whole tender. Such assessments should be conducted in a different order per member, to reduce the natural tendency to compare responses against those already considered. The panel can accept other relevant information as necessary, such as financial viability information and past performance, if this option is clarified in the evaluation.

3.5.3 Evaluation steps

The evaluation process can be broken down into the following sequential stages.

Stage I	Any late tenders must be identified, recorded and only accepted into evaluation in accordance with DPPM Chapter 5.5. A late tender should not be opened and accepted into evaluation, other than in accordance with the DPPM.
Stage II	All tenders, except any excluded at Stage I, should be subject to an initial screening to establish whether they comply with all (if any) mandatory or essential requirements, as well as minimum form and content requirements.
Stage III	This is the main event in the evaluation process during which tenders are assessed and ranked or scored against predetermined evaluation criteria.

Stage IV	The purpose of this stage is to identify the best overall value for money by bringing together the results of the Stage III evaluation and undertaking a comparative assessment of all tenders that summarises the costs/benefits/risks of each tender and identifies all key discriminators between competing offers.
Stage V	Although negotiation generally has the potential to improve procurement outcomes by reducing uncertainties identified during the tender evaluation process, it should not be required for most procurements unless the acquisition is complex and high risk.

3.5.4 *Evaluation criteria*

Evaluation criteria are used to assist in making an objective and fair procurement source decision and to ensure value for money. Evaluation criteria must be established and documented prior to seeking tenders.

Evaluation criteria will vary from one procurement to another. The following list provides a general guide to the sorts of criteria that likely to be suitable for many simple procurements:

- Total price and any discount arrangements.
- Quality of supply and technical suitability.
- Compliance with delivery requirements.
- Past performance of the supplier.
- Compliance with the conditions of contract.
- Supplier performance risk and capacity.

Many procurements contain a mix of goods and services with one or the other generally dominating. In the case of goods, the main emphasis in evaluation is generally on technical suitability, fitness for purpose, availability and price. In the case of services, the capacity and capability of the supplier and the quality and price of the service assume greater importance. It may also be necessary for service procurements to take into account a stronger emphasis on issues such as security (if facility access is required), intellectual property (if documentation is being generated under the contract), and possibly confidentiality and privacy issues as well.

In both cases, price is only one consideration. The primary concern and overall objective are to ensure that the selected supplier can complete the requirement to the desired standard in a way that balances the cost and benefits to be gained from buying the goods or services from that supplier (as compared to those available from alternative suppliers).

During the evaluation process it is critical that tenders are evaluated using the predetermined evaluation criteria. These criteria should not be altered once tenders have been sought.

3.5.5 *Evaluation methodologies*

The structure of a detailed tender evaluation, and how it is approached, will depend on the evaluation methodology (or model) that is selected for the procurement.

Least-cost-conforming tender method is a simple and common evaluation model used for many straightforward simple procurements. It effectively involves accepting the lowest-priced conforming tender (i.e. one that meets the specification requirements for the procurement).

Because this model essentially means competing over price, the tenderers should have the capability, capacity and financial capacity to successfully undertake the procurement. These factors should be ascertained before contract signing.

Comparative assessment and ranking method is considered suitable for simple procurements. It involves a comparative assessment and ranking of tenders, typically against a range of both price- and non-price-evaluation criteria. Under this model, evaluation criteria are effectively split between quality (technical suitability, fitness for purpose and availability) and price components of a tender. Tenders are evaluated both separately and wholly for quality before price is taken into account. Price should be considered last so as not to impact or influence other parts of the evaluation.

3.5.6 Whole-of-life costing

Because many complex engineering systems have expected lives of more than 25 years, the cost of ownership becomes a critical issue for the sustainability of the acquisition project.

An additional step that is sometimes required to achieve a true like-for-like pricing evaluation baseline involves calculation of a whole-of-life cost for each tender. This is also known as life-cycle costing or total procurement cost.

A whole-of-life cost may include the initial purchase price, installation costs, ongoing maintenance, cost of spares, licence fees, government-furnished material requirements or disposal costs. It may also take into account (where relevant) the timing of replacement of a product at end of its useful life.

3.5.7 Value for money assessment (VFM) and source selection

This final stage in the evaluation process involves a holistic comparative assessment of all tenders that summarises the costs, benefits and material risks of individual tenders and identifies all key areas of differences between tenders. Key inputs into this process (and the final VFM decision) are the results of the tender evaluation conducted during Stage III, coupled with exercise of professional judgement of the experienced tender evaluators who are undertaking the assessment. Upon completion of this assessment, an overall comparative ranking is assigned to the tenders.

It is common that after the comparative assessment, one technically superior tender requires a higher price even though a less-technically-capable solution exists at a lower price. The recommended approach to resolving this situation is to undertake a close and careful cost-benefit analysis of the two competing offers, using the lower-priced option as the standard against which the higher-priced option is fully justified and tested.

An appropriately detailed written record of the overall assessment must be maintained in all cases. This record should summarise the tender-evaluation process and contain details of all key decisions, with sufficient supporting reasons. This documentation is essential for ensuring that both the process and its outcome can be defended and can withstand challenge and scrutiny.

References

Eynard, B., Gallet, T., Nowak, P., Roucoules, L., 2004. UML based specifications of PDM product structure and workflow. Comput. Ind. 55 (3), 301–316.

International Council on Systems Engineering, 2011. Systems Engineering Handbook v. 3.2.2, INCOSE – TP – 2003 – 002 – 03.2.2, October, p. 373.

Noor, N.M.M., Mohemad, R., Man, M., Abdullah, A.I., 2008. Decision support for web-based prequalification tender management system in construction projects. In: Proceedings 2008 International Conference on Service Systems and Service Management. June 30–July 2, Melbourne, Australia.

Shen, H., Wall, B., Zaremba, M., Chen, Y.-L., Browne, J., 2004. Integration of business modelling methods for enterprise information system analysis and user requirements gathering. Comput. Ind. 54 (3), 307–323.

Systems design

4

Chapter Outline

In a broader sense, environmental factors include all aspects of the world that are outside the boundary of the complex engineering system. Figure 4-1 provides an overview of the factors that should be considered irrespective of the sector the product is designed for, by charting the assessment of the environmental factors on the product or service. The higher the rating (i.e. all points near the outer extent of the chart), the more positive is the factor for the systems support engineering requirements. In the design of a support solution, it is necessary to study the weak areas carefully and plan alternative actions to lessen their impact on plans within the solution.

Based on assessment by an independent body, the engineering solution can be positioned on the chart. After considering the impact of the previous factors, changes to the supporting analysis and its conclusions, the input should be adjusted until the support solution and plans match the characteristics of the service organisation:

- Planning and reporting.
- Maintenance plan.
- Supply support plan.

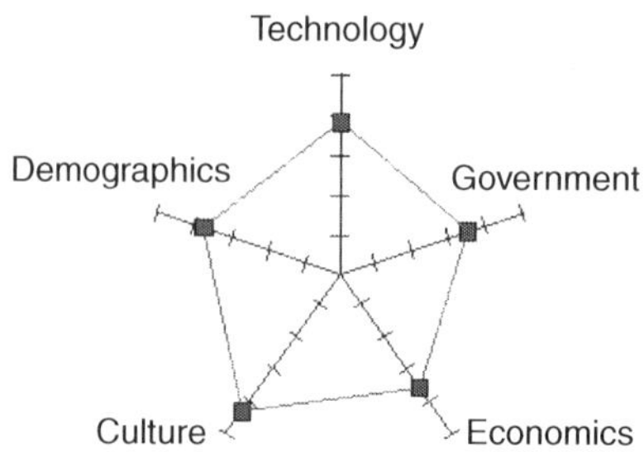

Figure 4-1 Environmental factors

- Personnel and training plan.
- PHST plan.
- Transition plan.
- Disposal plan.
- Software support plan.
- Post production support plan.
- System retirement plan.
- LSAR.
- FRACAS system.
- Environmental factors management plan including emission, noise and disposal (e.g. materials, liquids).

4.1 Systems design process

When designing complex engineering products, design harmonisation focuses heavily on functionality and performance. A well-balanced design is often the result of a well-organised decision-making process that includes all parts of the engineering organisation: designers, engineers, quality controllers, cost controllers, field operators etc. With team effort, new technologies can be considered and refined and risks and opportunities are better identified. Essential factors to be incorporated and harmonised during the design phase are performance, supportability and economic feasibility.

Remember that the engineering product is designed to satisfy the need of the customer. Stakeholders' involvement in the decision-making process is absolutely necessary. Fix a periodic meeting schedule with the customers to ensure that all parties are well informed and their opinions are heard. In addition, simultaneous, synchronised development of the mission system and the support system will help the newly changed system to be commissioned smoothly (Blanchard, 2008).

4.1.1 Protection of designs

The most important part of any design law is the definition of which product(s) or article(s) that are protectable. Distinction can be made on the basis of the significant elements of the different notions. For example, some definitions require industrial applicability; no handcrafted item can be the subject of protection; or only visually recognisable features are accepted. The UK's 1949 Registered Design Act (Intellectual

Property Office, 2014) is a good example of the old-fashioned industrial-ornamental design law philosophy. Only visible features that were also industrially applicable were eligible for protection. The current directive follows the new trend by giving a very broad definition:

> *A design is the appearance of the whole or a part of a product resulting from the features of, in particular the lines, contours, colours, shape, texture and/or materials of the product itself and/or its ornamentation.*

A product is defined as any industrial or handicraft item, including parts, packaging, get-up, and graphic symbols, but excludes computer programs (although computer graphics and icons aren't expressly excluded). No shape is registrable if its form is dictated solely by technical functions.

The broad definition of the design directive only applies to a non-laxative list of protected subject matter, which makes it very flexible. Industrial applicability is not a requirement, so handicraft items are also covered. Reproducibility is not required, although without technical ability to some kind of mass production an item may considered an article of fine arts; in this case it is protectible by other forms of legal protection (mainly copyright). The lack of industrial applicability requirement is also a characteristic of the widely respected definition, in design terms, of industrial system design. In other words, industrial design represents a creative activity that aims to define the design quality of industrially produced goods. These formal qualities are not only external features but are also basic structural and functional connections that create a coherent integrated system in the view of designers and users alike.

Another widely used definition of design describes it as any composition of lines or colours or any three-dimensional form, provided that such composition or form gives a special appearance to a product of industry or handicraft or can serve as a pattern for (such) a product. Because the protected subject matter pursuant to this definition is the individual appearance of the product, only an eye-appeal test is required. Design is defined as 'an aesthetic creation'. A technically functional element is protectable if it also has some aesthetic character and the same technical result could be reached by another shape.

An exception to design harmonisation by features concerns mechanical connections. In legal parlance, a design right shall not subsist in features of appearance of a product which must necessarily be reproduced in its exact form and dimensions in order to permit the product in which the design is incorporated or to which is applied to be mechanically connected to or placed in, around or against another product so that either product may perform its function.

This exclusion is effectively specified for the repair (or spare part) clause of the Directive. A spare part is a component of a complex product (protectable under the Directive); its purpose is to repair the original appearance of the product 'so as to restore its original appearance'. The necessity to restrict the legal protection of such parts originates from the fundamental idea of the EU to liberalise the movement of goods within the internal market. The protection law is implemented as a remuneration system for the use of protected designs, including the appropriate level of remuneration. Independent producers of spare parts may pay a reasonable remuneration to rightsholders.

It can be seen from the previous definitions and explanations that a number of challenges are inherent in design harmonisation, and that resolutions to these challenges have been evolving within legal systems for a number of years. In the restricted defence market, this issue is suppressed due to security and related protective measures that have already been established around the technologies that are applied to defence systems.

4.2 Systems modelling: functional, data, process

Systems modelling is the interdisciplinary study of the use of models to conceptualise and construct systems in engineering, business and IT development. A common type of systems modelling is functional modelling, which involves specific techniques such as the Functional Flow Block Diagram and IDEF0 (Figure 4.2-1). These models can be extended using functional decomposition and can be linked to requirements models for further systems partition. A model is a representation, often in the form of mathematical description, of the construction and workings of a system component. Models can be linked to represent their interaction. Simulation is the process of activating all models by introducing changes (e.g. time, current, personnel, forces) over a certain time period. The objective of modelling and simulation of systems is to replicate a system behaviour in an abstract environment where it can be studied, analysed and optimised before it is established in the real world. This approach is particularly powerful for very complex systems for which each individual element and how they function may be understood, but the behaviour of the entire system may not.

In practice, systems are modelled with different scopes and scales of complexity.

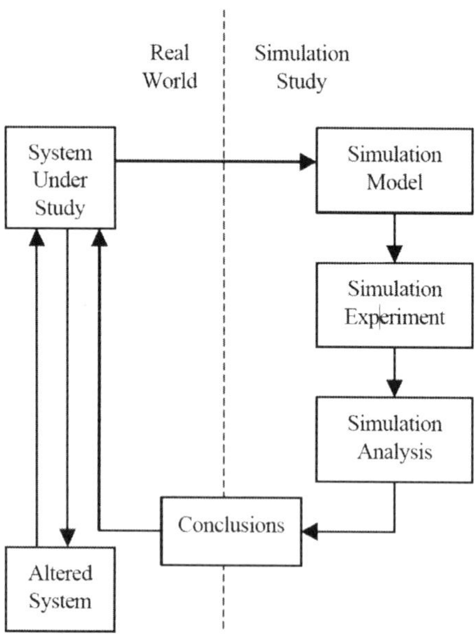

Figure 4.2-1 IDEF0 model

4.2.1 Functional modelling

Functional modelling in systems engineering and software engineering is a structured representation of functions (i.e. activities, actions, processes, operations) within the modelled system. A function model, similar to the activity model or process model, is a graphical representation of an enterprise's function within a defined scope (KBSI Inc, 1993). The purpose of the function model is to describe the functions and processes, assist with discovery of information needs, help identify opportunities and establish a basis for determining product and service costs. In systems engineering and software engineering, a function model is created from a functional modelling perspective. The functional perspective is one possible perspective in business process modelling; other perspectives include behavioural, organisational or informational.

A functional block diagram describes the functions and interrelationships of a system. It can picture:

- Functions of a system, depicted by blocks.
- Input and output elements of a block, pictured with lines.
- Relationships between functions.
- Functional sequences and paths for matter and/or signals.

A functional modelling perspective concentrates on describing the dynamic process. Its main concept is the process, which could be a function, transformation, activity, action, task etc. A well-known example of a modelling language that employs this perspective is data-flow diagrams (Figure 4.2.1-1) (Harel, 1987). This perspective uses four symbols:

- *Process*, which illustrates transformation from input to output.
- *Store*, which refers to data collection or some sort of material.
- *Flow*, which means movement of data or material in the process.
- *External Entity*, which interacts with the modelled system despite being outside of it.

A process can be represented as a network of these symbols. A graphical representation of this decomposed process is called a data flow diagram (DFD).

4.2.2 Functional decomposition

Functional decomposition refers broadly to the process of resolving a functional relationship into its constituent parts in such a way that the original function can be reconstructed from those parts. In general, this process of decomposition is undertaken either for the purpose of gaining insight into the identity of the constituent components or for the purpose of obtaining a compressed representation of the global function—a task that is feasible only when the constituent processes possess a certain level of modularity. In functional decomposition of engineering systems, which is a method for analysing engineered systems, the basic idea is to try to divide a system in such a way that each block of the block diagram can be described without using the words 'and' or 'or'. This exercise forces each part of the system to be assigned a pure function. When a system is composed of pure functions, they can be reused or replaced. A usual side-effect of such composition is that the interfaces between blocks become simple and generic. Because the interfaces usually become simple, it is easier to replace a pure function with a related, similar function. As we have previously stated, a process can be represented as a network of these symbols. Such a symbol network of a decomposed process is called a data flow diagram or DFD.

A functional flow block diagram (FFBD) is a multi-tier, time-sequenced, step-by-step flow diagram of a system's functional flow. FFBD notation, which was developed in the 1950s, is widely used in classical systems engineering. FFBDs are one of the classic business-process-modelling methodologies, along with flow charts, data flow diagrams, control flow diagrams, Gantt charts, PERT diagrams and IDEF. FFBDs can be developed in a series of levels. FFBDs show the same tasks identified through functional decomposition and display them in terms of logical, sequential relationships. Each block in the first-level diagram can then be expanded to a series of functions, as shown in the second-level diagram for 'perform mission operations'. Note that the diagram shows both input (transfer to operational orbit) and output (transfer to space transportation system orbit), which thereby initiates the interface identification and control process. Each block in the second-level diagram can be progressively developed into a series of functions, as shown in the third-level diagram in Figure 4.2.1-1.

The FFBD also incorporates alternate and contingency operations, which improves the probability of mission success. The flow diagram provides an understanding of total operation of the system, serves as a basis for development of operational and contingency procedures and pinpoints areas where changes in operational procedures could simplify the overall system operation. In certain cases, alternate FFBDs may be used to represent various means of satisfying a particular function until data are acquired, which permits selection among the alternatives.

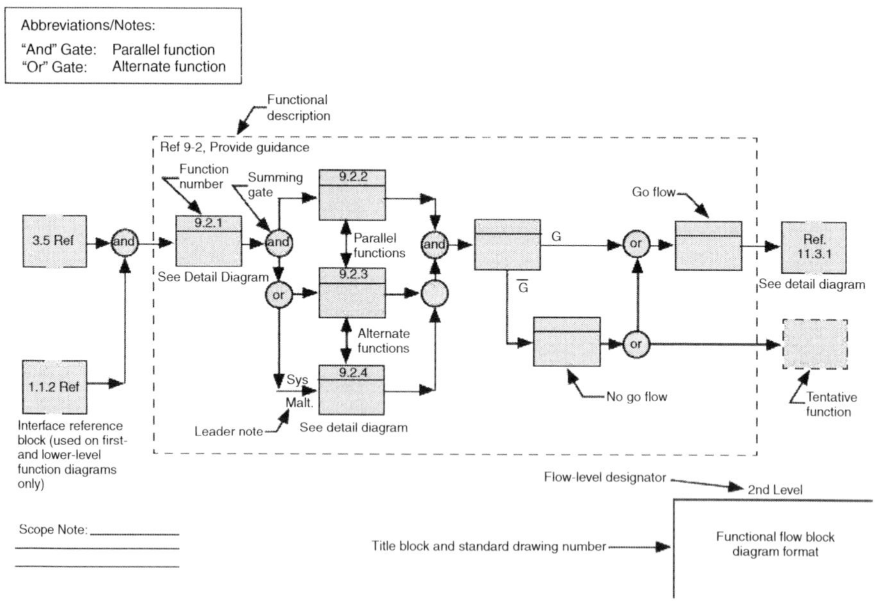

Figure 4.2.1-1 Data flow diagram

4.2.3 Data modelling

A data model is a description of the objects represented by a computer system, together with their properties and relationships; typically, these are real-world objects such as products, suppliers, customers and orders. Data models often aid communication between the business people defining the requirements for a computer system and the technical people defining the design in response to those requirements; they are used to show the data needed and created by business processes. Data models are specified in a data-modelling notation, which is often graphical in form. A data model can be referred to as a data structure, especially in the context of programming languages. Data models are often complemented by function models, especially in the context of enterprise models.

4.2.4 Process modelling

A process model exemplifies or represents processes of the same nature, which are classified together. Because a process model is presented at the type level, each individual process within the model is an instantiation. Thus, many instantiations can be contained within the same process model, which can be used repeatedly for the development of many applications. One possible use of a process model is to prescribe how things must be done; this prescriptions stands in contrast to the process itself, which is what really happens. In short, a process model is roughly an anticipation of what a process will look like. What the process will in fact be is then determined during actual system development.

4.3 Design for sustainability

When considering engineering change to an existing system, many factors must be considered. In most existing operational systems, performance must be assessed prior to any material change. The result of the assessment can be used as a benchmark to measure success of a change and as input data to future additional enhancements.

4.3.1 Built-in capabilities

After performance baseline data have been gathered, a number of built-in capabilities within the system to be changed must be considered during the design stage. They are:

- Built-in test (BIT), meaning the capability to enable continuous monitoring, maintenance and upgrades.
- Diagnostic capability to enable fault analysis.
- Commonality of LRUs/parts to maximise the stock availability. This will have an impact on system availability, especially in critical mission situations.
- User skill sets, particularly when developing the human-machine interface.
- Manning and skill-set requirements when the system is operated after delivery.
- Levels of maintenance, which determine how quickly the system can be returned to service.
- Storage/retrieval of the system design data needs, which must be considered for future changes and baseline data management.

4.3.2 Detailed design activities

Several detailed design activities are essential for the implementation of engineering changes. In fact, engineering change by itself is a support activity. Thus, it is in the interest of the engineering-change designer to maintain or, if possible, improve the supportability characteristics of the design solution. Doing so may mean making significant enhancements to some of the existing installations that are not initially considered to be changed. Always test and evaluate the design solution to ensure that all supportability requirements have been achieved.

In addition, always design the solution for reliability. Design for Reliability is an estimated parameter, but the baseline data should come from the in-service performance of existing or similar equipment (such as MTBF) with due consideration of the changes that will be made during the engineering-change process. Two other factors should also be considered in relation to design for reliability:

- Item selection.
- Redundancy.

Similarly, design for maintainability is required to ensure that the same or better maintenance process will reduce MTTR and thus improve the availability of the system.

In designing for maintainability, the following factors are considered:

- Reliability-centred maintenance.
- Online test and diagnostics.
- Fault detection.
- Design for availability.
- Allocation of support requirements.
- Reliability allocation, modelling and prediction.
- Operational availability.
- Maintainability allocation, modelling and prediction.

In any complex engineering system, there is always some involvement of humans (e.g. operator, decision maker such as a mission-system supervisor, inspector, technician). A complete engineering solution should include sufficient assessment of the effect on the humans in the system. Design for human factors is not just designing the HMI; it is about factoring in all scenarios in which a human entity will be engaged. Estimation of the reliability of the human in an engineering system is difficult but it is an essential step towards high reliability. Analyses that are important to assist in the design for human factors include:

- Task analysis (operator and maintainer).
- Trade-off studies (e.g. buying/making decisions).

Design for supportability takes a broader view. Review supportability requirements at each design review (formal and internal), including but not limited to the following factors:

- Logistics support analysis (LSA).
- Minimisation of needs for in-service support engineering.
- Level of repair analysis (LORA).

- Life cycle cost (LCC) analysis.
- Standardisation.
- Maintenance-concept development and optimisation.
- Cost-effectiveness analysis.
- Optimal system/equipment packaging design.
- Support-recourse requirements.
- Personnel requirements.

The following techniques can be applied to assist in the analyses during an engineering-design process:

- Comparative analysis.
- Functional analysis.
- Technology road-mapping.
- Modelling and analysis that includes the support system and environment.
- Tools (FMECA, reliability modelling, spares modelling, task analysis etc.).
- Simulation.
- Sensitivity and risk analyses.

4.4 Tool design

Tools are designed and manufactured to support the manufacturing process. The purposes of tool design are to:

- Reduce the overall cost of manufacturing a product by producing acceptable parts at lowest cost.
- Increase the production rate by designing tools that will produce parts as quickly as possible.
- Maintain quality by designing tools that will consistently produce parts with the required precision.
- Reduce the cost of special tooling by making every design as cost-effective and efficient as possible.
- Design tools that will be safe and easy to operate.

Some examples of manufacturing tools are:

- Cutting tools, tool holders and cutting fluids.
- Machine tools.
- Jigs and fixtures.
- Gauges and measuring instruments.
- Dies for sheet metal cutting and forming.
- Dies for forging, cold finishing and extrusion.
- Fixtures for welding, riveting and other uses.
- Mechanical fasteners.

The manufacturing tools can be a significant part in the overall product cost process. In most cases specific to the design, tools must not wear out and must withstand high loads and temperatures. Figure 4.4-1 shows an example of a high-value tool for large assets. During the design process, attention must be paid to how the product is going to be made and what design tools are required.

Figure 4.4-1 A high-value tool

The process for tool design is:

1. Statement and analysis of the problem.
   a. Problem without tooling.
   b. What the tool is supposed to do?
   c. Bottleneck in assembly.
2. Analysis of the requirements.
   a. Must perform certain functions.
   b. Must meet certain minimum precision requirements.
   c. Must keep the cost to a minimum.
   d. Must be available when the production schedule requires it.
   e. Must be operated safely.
   f. Must meet other requirements such as adaptability to the machine tool etc.
3. Development of initial ideas.
4. Development of design alternatives.
5. Finalisation of design ideas.

4.5 Design verification

Requirements for verification and validation depend upon the sophistication of the engineering solution. The validation process is concerned with checking the requirements for omissions, conflicts and ambiguities and for ensuring that the requirements follow quality standards. This should be undertaken by a multi-disciplinary team that includes members of the supportability engineering team.

The term 'verification' is in fact a lower-level activity (closer to product). The engineer should ask: Am I building the product correctly? The review of interim work steps and interim deliverables during a project helps to ensure they are acceptable. It also determines whether the system is consistent, adheres to standards, uses reliable techniques and prudent practices, and performs the selected functions in the correct manner. Always throw up the question: Am I accessing the data 'right' [in the right place; in the right way]? Parts of the work that is to be performed during development will be on key artefacts such as walkthroughs, reviews and inspections, mentor feedback, training, checklists and standards. During the implementation process, demonstration of consistency, completeness and correctness of the software at each stage and between each stage of the development life cycle will always minimise the chance of error.

On the contrary, validation is a high-level activity (closer to design). The engineer should ask: Did I build the right product? In product validation, the system designer should determine if the system complies with the requirements, performs functions for which it is intended and meets the organisation's goals and user needs. Always ask: Am I accessing the right data (in terms of the data required to satisfy the requirement)? The validation process is performed after a work product has been produced, against established criteria which ensure that the product integrates correctly into the environment. The aim is to determinate the correctness of the final software product from a development project with respect to user needs and requirements.

For complex engineering products, verification of requirements should include supportability requirements. The LSA program develops quantitative requirements for mission system supportability and the supportability-related requirements of the enabling support systems. Quantitative requirements, once defined, are used in developing measurable supportability-design requirements and support-system requirements for specification documents, as well as for verification and validation activities.

The following techniques are commonly used in verification and validation:

- Methodologies: test and/or simulation.
- Reliability and maintainability analysis.
- Reliability testing.
- Maintainability demonstration.

4.6 Design freeze

Freezes play a major role during product development. Many companies use high-level stage-gateway processes for new product development (NPD) in which freezes mark the end-point of a development stage. For example, 'specification freeze' defines the set of requirements on which the entire design will be based and 'design freeze' describes the end-point of the design phase at which a technical product description is handed over to production. Stage-gateway processes usually depict a single point for design freeze (Figure 4.6-1). However, some systems, parts, features or parameters must be frozen prior to the official design freeze; dependencies between parameters require early definition and long-lead-time items must be completed ahead of time.

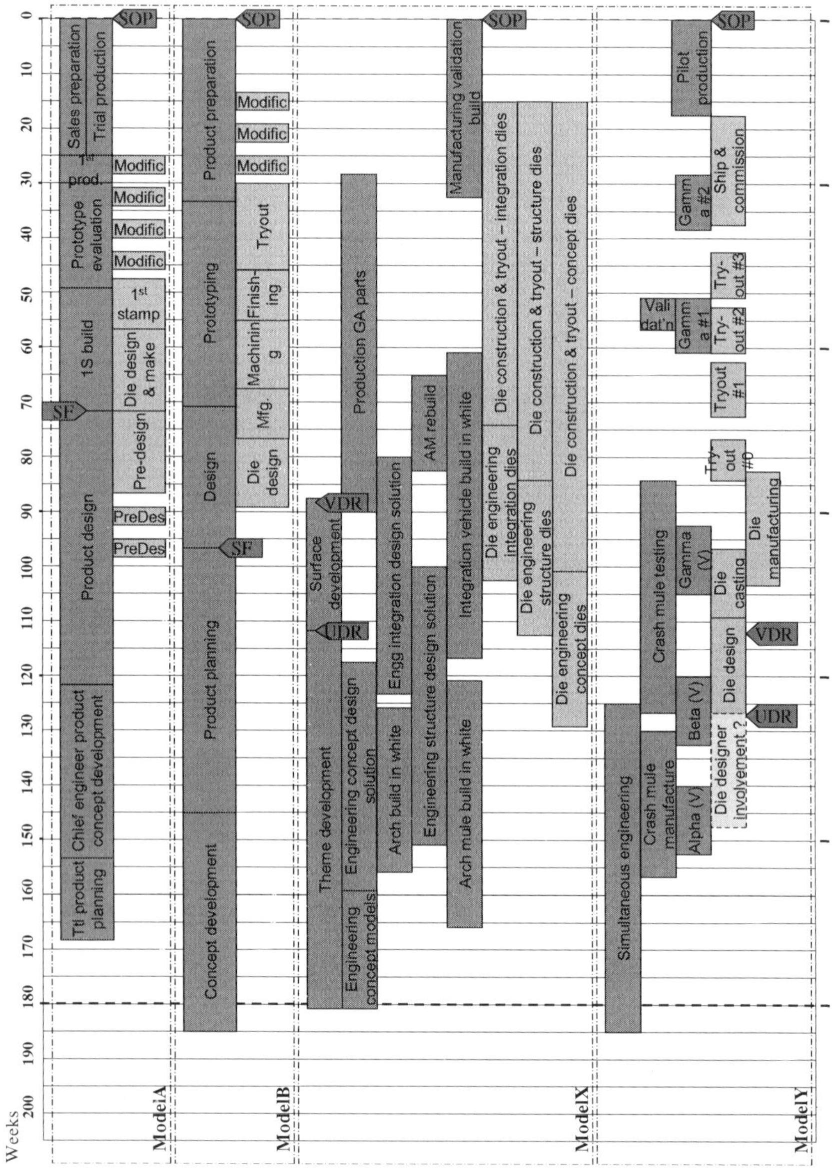

Figure 4.6-1 Design freeze in a vehicle-development plan

Freezes occur on three levels of detail:

- *Product freezes*. The term 'design freeze' is most commonly used to denote the entire product design. A product freeze is a single point in time that marks the end of the design phase.
- *Part freezes*. This term refers to single parts, or groups of parts, that are frozen at the same time (as opposed to a 'system freeze', which refers to the entire system). In the automotive industry, a typical freeze of a group of parts is called a 'style freeze'. By this point in time, the interior design of a car is frozen and both the shape and the available room for all parts in the passenger cabin have been defined. Of course, this type of freeze is done in conjunction with the engineers who have been responsible for the separate parts. Individual parts are frozen at different times, in other words, to allow for design continuation of dependent components. Part freezes are frequently driven by lead time. For example, in one car interior design process the airbag was the longest lead-time item. Therefore, it had to be defined within a few months of starting work on the car and thus fixed many of the airbag's interface parameters.
- *Parameter freezes*. Parts are not usually frozen at a single point in time. Instead, parameters, features and the interfaces to other parts are frozen individually before the whole part is approved. Key parameters that determine the performance, function and manufacturability of the whole part are usually set first. For example, while the material for a part may be decided early on to allow for procurement, the exact part shape is set later. Even the shape can be set in stages, as long as 'metal-off' is possible (i.e. as long as the part can be machined into its final form). Parameter freezes structure the design process. The dependency of key parameters sets a type of internal logic of the product, which governs process planning as well as decision ordering throughout the entire production process. Without defining key parameters, dependent decisions cannot be finalised.

One of the objectives of design freezes is to prevent further engineering changes. For example, freezes avoid new technologies that can be implemented in the next product generation. However, it may be necessary to include other changes such as safety issues, problem corrections or new customer requests. Changes that must be implemented after freeze may be more costly if tooling and so forth are already in place. Quality control norms like ISO 9000 require a freeze point for change control, in order to distinguish between the design phase and the subsequent change-implementation phase.

Product parts are rarely totally fluid until they are suddenly fixed at a freeze point. Instead they are often 'chilled', which makes changes less likely and more costly until the component reaches a point at which it will be only modified if the integrity of the product is jeopardised. A 'chill' in this context signifies that the design of a part has been completed. When changes must be carried out within a product architecture in which some parts have already been chilled or frozen, a preferred implementation avoids changes to these parts. Unfortunately, the complexity and scale of dependencies between components often make it difficult to correctly predict the knock-on effects of changes.

Different parts evolve at different rates. In Figure 4.6-2, the interface between parts A and B is frozen early so that the design of part B can begin. This figure also shows part lead-times: although part C depends on part B, it has a longer lead time and therefore needs to be frozen first. Figure 4.6-2 indicates part lead-times and design times that are the result of dependencies in the system. Preliminary information can be used to begin dependent part designs before an upstream part has been completed. Freezing some of this information and using it as a basis for further design reduces the risk of rework.

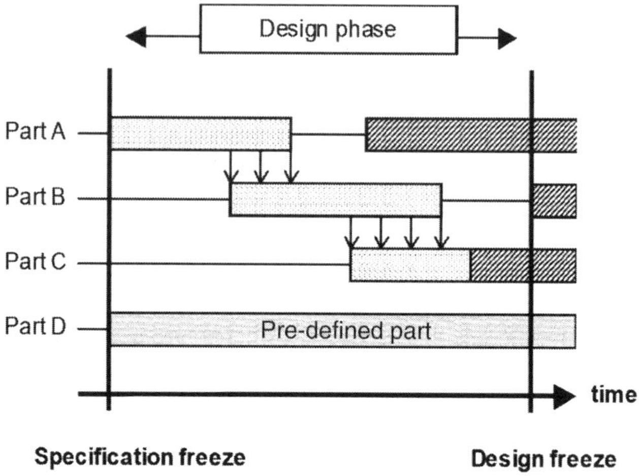

Figure 4.6-2 Design phases

Figure 4.6-2 also indicates the use of a predefined part, for example a standard component that is bought in or a legacy part or platform part that is carried over from another product. The use of predefined parts gives a starting point for the design process, reduces the risk in performance and design and may lead to economies of scale. It reduces the workload on the designer unless the effort in integrating the part is higher than the savings it may achieve. Predefined parts, such as off-the-shelf parts, are frozen externally and the designer only has limited influence on the design. Although a standard part might be replaced by another, a change to a legacy or platform part has more severe consequences; specifically, the cost benefits of using a predefined part would be lost. Predefined parts can be considered frozen from the start of the design process.

The timing of freezes is often beyond the control of a company, as it is heavily dictated by lead times. Decisions that refer to key parameters of the entire system need to be made early to reduce the risk of rework. Fortunately, companies have some flexibility over when they freeze a part.

4.7 Maintenance requirement determination

The development of maintenance policy for technical equipment involves the systematic application of a set of defined processes. These processes are collectively referred to as maintenance requirements determination (MRD).

The MRD process involves:

- *Functional analysis*. This form of analysis is based on the definition of system operational requirements and the system or equipment maintenance concept and is used as the basis of detailed design. MRD requires the results of functional analysis in order to identify failure modes, causes and effects, and associated criticality using failure modes, effect and criticality analysis (FMECA).

- *Maintenance determination.* Once the failure modes and the effect of failure have been determined, they are used as inputs for determining the corrective and preventative maintenance requirements. The corrective maintenance determination is focused on identifying the necessary repair actions required to return an item to serviceability. The preventative maintenance outcome is identified by the application of reliability centred maintenance (RCM)-based methodologies.
- *Maintenance task analysis (MTA).* After identifying the corrective and preventative maintenance requirements, the logistics resources necessary to support these requirements must be identified through the process of MTA. MTA will identify resources; for example:
  - Repairable and breakdown spares.
  - Trade skills and training.
  - Packaging handling and transport.
  - Procedures required to perform the task.
  - Facilities.
  - Support and test equipment.
- *Level of repair analysis (LORA).* This consists of specialised models that look at the life-cycle cost (LCC) of performing maintenance tasks under different support scenarios. Used to determine the most cost-effective maintenance policy for a system or equipment. Maintenance policy must be subjected to appropriate LORA to ensure that defence resources are being efficiently and effectively utilised. Many of the commercially available LORA models are packaged with functions such as spares optimisation and LCC analysis (LCCA).
- *Logistic support analysis record.* The results of MRD must be documented in a manner that will enable the data to be used and updated throughout the material life cycle. Because MRD and LSA require similar inputs and share many common processes, the MRD results can be readily stored in a logistic support analysis record (LSAR) that has been modified with unique ADF tables.
- *Promulgation.* The promulgation of maintenance requirements occurs through the issue of maintenance manuals and servicing schedules.
- *Performance monitoring and analysis.* Performance monitoring, which involves the proactive monitoring of the maintenance program, can take many forms and utilise a variety of tools. The main aim is to identify developing trends or degraded performance. Such signs indicate the necessity of adjustment or review of one or more elements of the maintenance program, or of logistics support.

4.8 Reliability-Centred maintenance (RCM)

RCM is a process used to determine what must be done to ensure that any physical asset continues to do what its users want it to do in its present operating context. Application of RCM techniques promotes safety and maximum system availability at minimal cost over the material life cycle (Takata et al., 2004). This is typically achieved through:

- Ensuring that existing tasks are technically correct (applicable) with respect to a known failure mode.
- Reviewing current task intervals to ensure they are effective in reducing the probability of failure and that the costs of maintenance are at an acceptable level.
- Assessing whether there are any new failures that need to be addressed.

RCM has seven basic steps, represented by the following questions:

- What are the functions and associated performance standards of the asset in its present operating context?
- In what ways does it fail to fulfil its functions?
- What causes each functional failure?
- What happens when each failure occurs?
- In what way does each failure matter?
- What can be done to predict or prevent each failure?
- What should be done is a suitable proactive task cannot be found?

4.8.1 Selecting items for analysis

A complex weapons system contains a vast number of parts and components, all of which have specific functions. Any one of these items can be expected to fail at one time or another and some of the failures will have more serious consequences than others. Certain kinds of failures threaten safety, whereas others may have direct effects on operating capability. However, there are a large number of items for which failure has no immediate impact on the application as a whole. These failures are simply corrected soon after they occur, and the consequences are limited to repair costs.

Item selection, which in RCM terms is called significant item (SI) selection, is necessary because applying the RCM analysis process to every component of a weapons system is neither cost-effective nor essential to ensure reliability. To maintain the RCM analysis at a workable level, the analysis should be performed at the highest possible level. Selection of all line-replaceable units (LRU) and shop-replaceable units (SRU) as SIs is both unnecessary and costly. Some items, particularly those in the avionics category, do not benefit greatly from an RCM analysis. SIs should be selected at the LRU level only if that LRU has a significant function.

4.8.2 Failure modes and effects analysis (FMEA)

Once the SIs have been identified, FMEA data for each failure mode is used during the RCM logic process to identify if an applicable and effective task can be performed. Failure-mode data for new equipment can be obtained or developed from FMEA worksheets (Figure 4.8.2-1).

4.8.3 Determining failure-mode significance

Once FMEAs for each item have been established, the RCM process begins with a review of each of the failure modes that have been identified, to see which ones are significant for RCM purposes. Applying RCM to every failure mode may unnecessarily consume analysis resources as some failures are not critical but are low-cost and easy to repair. Therefore a ‘run to failure’ strategy may be readily apparent, which would eliminate the need to go through the RCM logic process. The analyst applies the questions in Figure 4.8.2-1, external of the RCM logic process. If a YES answer is the outcome of any of the four questions, the RCM process should be initiated

FMEA - Overheat Detectors

Item Name:	Detector 1-4		Part #:	150D01-150		LCN/ALC:	A04/00 - AOA/00	
FMI	FUNCTION	FUNCTIONAL FAILURE	FAILURE MODE	FAILURE CAUSE	FAILURE EFFECTS			DETECTABILITY
					LOCAL	NEXT HIGHER	END EFFECT	
F1A1	To sense an Overheat condition within the Jetpipe area defined as >150°±10	Failure to sense an Overheat condition within the Jetpipe area defined as >150°±10.	Overheat Detector switch contacts fail to close. (No indication of Overheat condition).	Surface damage and corrosion. Moisture Ingress.	Overheat Detector fails to sense an Overheat condition. No warning signal sent to CWS.	Overheating will go undetected and corrective action will not be initiated.	Combined with another failure - Engine faults, cracking of Jetpipe or Fuselage Structure.	Nil
F1B1			Overheat Detector contacts close at < 150°. (Early closure of switch contacts or false indication).	Surface damage and corrosion. Moisture Ingress.	CWP displays false 'JPOHT' warning.	Immediate Mission cancellation.	N/A	Nil
F1C1			Overheat Detector contacts close at >160°. (Delayed indication of Overheat condition).	Surface damage and corrosion. Moisture Ingress.	Overheat Detector contacts close at >160°.	Delayed initiation of CWS 'JPOHT' warning. Delayed corrective action.	Combined with another failure - Engine faults, cracking of Jetpipe or Fuselage Structure.	Nil

NOTE: The Jetpipe Overheat System consists of four switches connected in parallel. As only one is required to sense an overheat condition a level of redundancy is provided.

Figure 4.8.2-1 Example FMEA worksheet

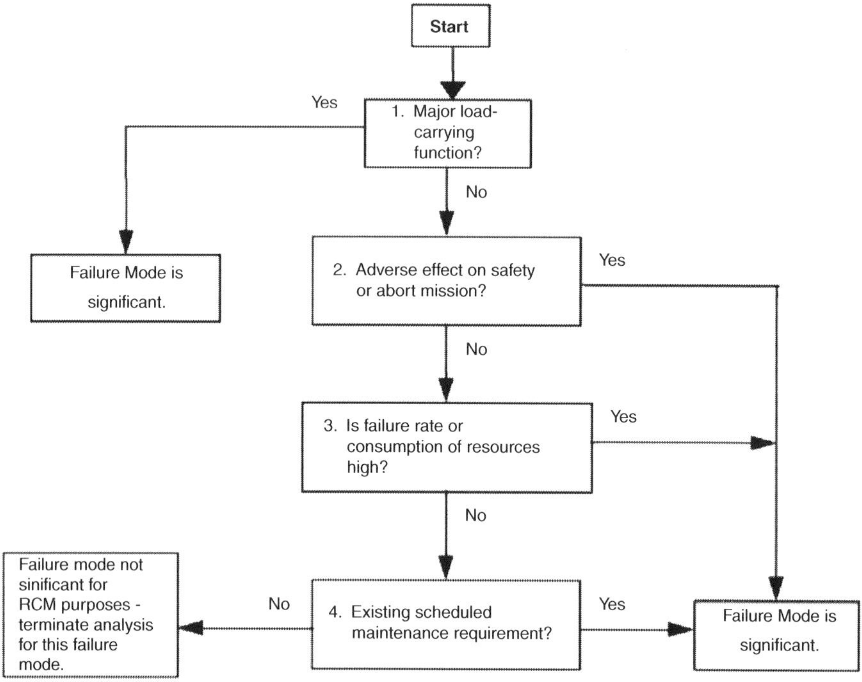

Figure 4.8.3-1 Failure mode significance determination

within the application in order to access the preventive maintenance requirements. Conversely, if a NO answer is the outcome of all four questions, no further RCM analysis is required for the failure mode under consideration. In such cases, the analyst should provide a summary of the questions and answers in Figure 4.8.3-1 in the RCM reasoning narrative.

4.8.4 Functional item analysis

Each failure consequences branch requires the application of different criteria during the task-evaluation process that is appropriate to the consequences of failure. After the appropriate failure consequences have been determined, answers to the remaining questions in the functional logic determine whether a preventative maintenance task is justified, according to a set of applicability and effectiveness criteria. The criteria are unique to each type of task. If the criteria are met, the appropriate task will be developed to address the failure mode.

The first question in Figure 4.8.4-1 determines whether the functional logic questions or the structural questions should be asked. The next questions determine the failure consequence for each failure mode and identify which failure consequence branch to follow during task evaluation.

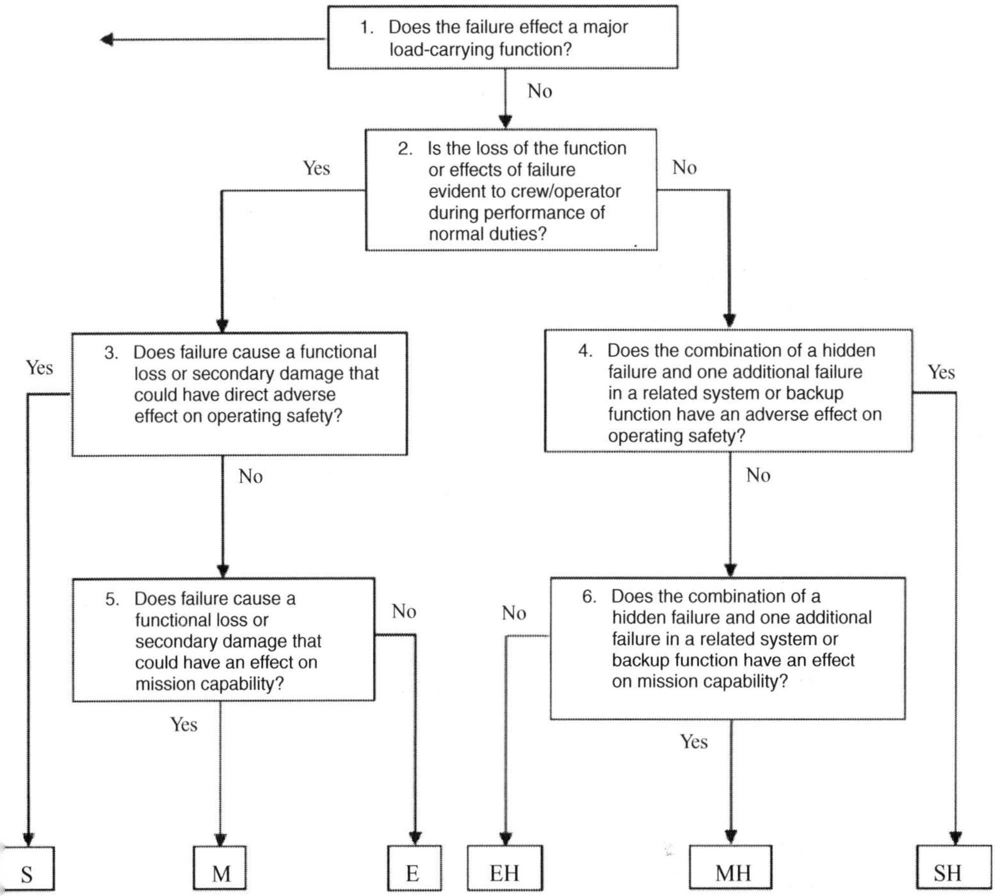

Figure 4.8.4-1 Consequences of failure determination

4.8.5 Determining failure consequences

The primary concern with failure is its consequences. These consequences may range from the modest cost of replacing a failed component to the possible destruction of a piece of equipment or the loss of life. Thus, all maintenance determination, including the need for redesign, is dictated by the nature of the consequences of failure. A preventative maintenance program is therefore based on the precept that 'the consequences of a failure determine the priority of the maintenance activities or design improvements required to prevent its occurrence'. The failure consequences determination questions are shown in Figure 4.8.3-1. The RCM-derived failure consequences, which are derived using Figure 3, are interpreted as a measure of the criticality of a failure mode, item or system. The criticality ratings from highest to lowest are safety (S), safety hidden (SH), mission (M), mission hidden (MH), economic (E), and economic hidden (EH).

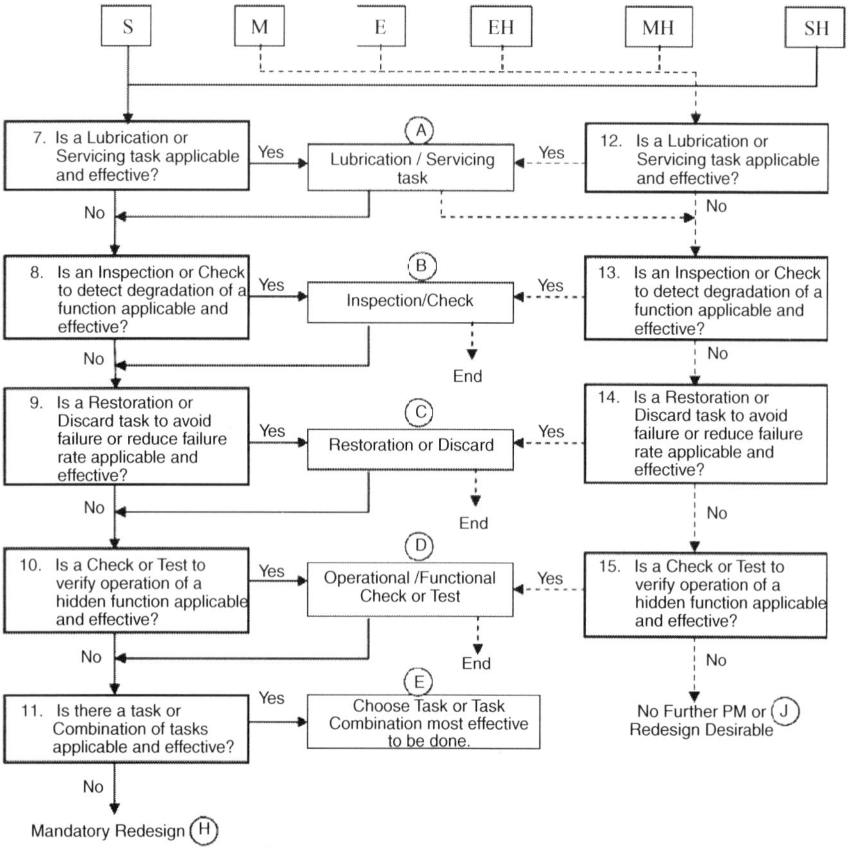

Figure 4.8.6-1 Task applicability and effectiveness determination

4.8.6 Task evaluation

The RCM task evaluation questions, shown in Figure 4.8.6-1, require a task to meet specific applicability and effectiveness criteria to be considered acceptable. An applicability criterion determines whether the type of task is appropriate for preventing the failure and therefore depends upon the failure characteristics of an item. The effectiveness criterion determines whether the task can be performed at a practical interval that will either reduce the probability of failure to an acceptable level (when safety is a concern), or be more cost-effective than allowing the failure to occur (when safety is not a concern).

4.9 Certification

An authorised engineering organisation (AEO) is a company that is authorised by the appropriate authority within a country to provide engineering services to support certain type of engineering assets. AEO certification is a significant achievement of a company, as it acknowledges the company's compliance with the technical regulatory framework in which the asset is to operate and comply. For example, the RAAF Technical Airworthiness Regulatory Framework is the Australian defence compliance authority. For military systems, the AEO certification will enable the company to provide critical engineering services to the Australian Defence Force. Becoming an AEO is an important component of continuous support of a platform in a country. It will normally involve an integration of a range of support capabilities and is the result of excellent teamwork and collaboration among the companies in place.

Integrated with the AEO is the ownership of design authority on the platform. The role of the design authority is to:

- Develop, maintain and communicate a set of technical standards that support the platform's activities.
- Participate in pre-business-case product specifications and high-level project-design activities to ensure alignment with standards.
- Approve compliance with or variance from the standards for all engineering projects, across all departments.

The adoption and use of technology innovations is a central theme of improving system capabilities. Technology innovations encompass new scientific inventions, computers, and other information and communication technologies. Much of the early work on technology innovation assumed that the technology was fixed whereas the job of designers was to adapt or modify the technology parameters to suit an application. This approach led to a disparity between the intentions of those designing and implementing the technologies and their certification. Even today, there are conflicting interpretations of this disparity. Some have seen the use of technology that is not faithful to its designers' intentions as dysfunctional or the result of poor alignment that prevents users from gaining the maximum benefits from the technology. Others have seen it as an integral aspect of use and hence essential to the requirements for subsequent engineering changes.

Studies of technology in history and sociology state that a technology can always be other than it is. Social constructivist theories suggest that technology is shaped by a range of influences as a part of its development and then reshaped in use. Further, the same technology may be used in different ways in different settings, leading to different outcomes. This reflects interpretive flexibility (i.e. that any object, institution or process may mean different things to different people). Therefore, not only may a technology be adapted by different users but the same technology may also be interpreted in different ways. Consequently there is no linear, deterministic path between the design of a technology, its use and its impacts: users appropriate a technology innovation. Hence, the concept of operational envelopes in which the technology is designed and used in different systems (i.e. applications) should be made clear at the time a technology is adopted.

When a technology is being developed after intensive research, there are always expectations from different stakeholders in the project. When the technology is transformed into a system, the engineering design process is heavily influenced by the anticipated functionality as understood from controlled experiments and operating environments. There is often a gap between the designed and actual effect of technology in use. Technology as designed represents the views of designers, marketers and users of the requirements captured during design, whereas the technology in use represents the users' needs that are expressed through their intended actions as they operate the system in which the technology is implemented. The features, functions and theory of use of technology as designed are shaped by a range of influences (social, political, economic and technical) during the design process. However, the technology has not yet reached its final, stable state. The designed use of the technology should be completed by users who configure and customise the technology, who construct workarounds to circumvent the limitations of the embodied theory and who use the features and functions in unanticipated ways and for unanticipated purposes or activities.

The certification of a system thus represents the completion of the design process. In other words, the initial design process has resulted in the development of an engineering system that is now adapted (with changed features or functions) and certified within the capabilities provided by the technology as designed. If user appropriation of a technology innovation is viewed as the completion of the innovation design process, then understanding the influences on and outcomes of the appropriation process provides the basis for improving design.

A key consideration of certification is safety. Safety management system (SMS) refers to a comprehensive business management system that is designed to manage occupational safety and health elements in the workplace. An SMS for an in-service system is intended to be used as a framework to allow an organisation to meet its legal obligations under occupational health and safety laws while the system is operational. The structure of an SMS is not of itself a legal requirement but it is an extremely effective tool with which to organise the myriad aspects of occupational safety and health (OSH) that can exist within an organisation; often it is used to meet standards that exceed the minimum legal requirement.

An SMS is only as good as its implementation. Effective safety management means that an organisation needs to ensure that it is looking at all the risks within that organisation as a single system, rather than having multiple and competing silos of operation. If safety is not seen holistically, fractured views can interfere with the prioritisation of improvements or even result in safety issues being missed.

References

Blanchard, B.S., 2008. System Engineering Management, fourth ed. ISBN 978-0-470-16735-9, pp. 560.

Harel, D., 1987. Statecharts: a visual formalism for complex systems. Sci. Comput. Program. 8 (3), 231–274.

Intellectual Property Office, UK, 2014. Registered Designs Act An unofficial consolidated version of The Registered Designs Act 1949 produced by the Trade Marks and Designs Directorate, October, 2014, p. 68.

KBSI Inc, 1993. Announcing the Standard for Integration Definition for Function Modeling (IDEF0), Processing Standards Publication 183, December 21. http://www.idef.com/pdf/idef0.pdf.

Takata, S., Kimura, F., van Houten, F.J.A.M., Westkamper, E., Shpitaini, M., Ceglarek, D., Lee, J., 2004. Maintenance: changing role in life cycle management. CIRP Ann. Manuf. Tech. 53 (2), 643–655.

Management of engineering-design changes

5

Chapter Outline

Engineering activity does not end after introduction of the asset into service. Engineering remains an integral part of system support throughout the asset's operational life (Johnstone et al., 2009). Reasons for engineering during the life cycle of the asset include:

- Replacement of faulty or obsolete parts with newer parts.
- Development of nonstandard repair procedure.
- Adding new equipment to the system (i.e. upgrades).
- Extension of asset system life or operations beyond the purpose for which it was designed.

Management of engineering design changes is the action of providing engineering support to ensure that the product works the way it was designed and performs its defined role. It is also known as in-service engineering. Other services include maintaining product integrity (i.e. system safety/airworthiness), performing configuration management of the product design-and-build standards, supporting product availability and supporting product cost effectiveness.

Components, particularly commercial-off-the-shelf (COTS) components, are expected to become obsolete well before the end of the operational life of the system and to no longer be available. This means that an alternative replacement component must be installed and will require an engineering analysis, test and verification. Most commonly used COTS components are computer hardware and software. In-service damage, for example crash or battle damage, may require a customised repair procedure that must be supported by engineering analysis. A common in-service engineering activity is system upgrade by adding or replacing components with the objective of increasing system effectiveness. Examples are installing a new propulsion system, a new weapons system, additional fuel tanks, new avionics and so forth.

Engineering changes require approval by the design authority, which has already defined the product design standard during development. The product is manufactured to this design standard, with deviations/concessions. The in-service engineering authority manages the in-service build and design standards.

5.1 In-service phase

The main characteristics of the in-service phase are that development funding has been committed and expended; product configuration has been defined; costs associated with significant design change are large; and lots of little issues and the odd big one still must be resolved. The overall goal in this phase is to support the product's use in its defined role.

The essence of in-service engineering is the requirement that the asset remains functional and is being utilised while the service is performed. Because significant constraints are in force, preplanning is absolutely essential to reduce the risk of rendering the asset unusable during the process.

Maintenance requirements are subject to continual evolution throughout the service life of a system. Figure 5.1-1 shows a typical cycle of in-service maintenance requirements for aircraft. Routine monitoring and analysis is the normal approach to in-service support. However, typically there are points within an asset's in-service life at which a partial or full maintenance-policy review would improve the asset's safety, availability and cost of ownership. The application of reliability-centred maintenance (RCM) to in-service equipment can determine whether the preventative maintenance being performed is valid. For example, a 26% reduction in tasks can be achieved (Figure 5.1-2). Applying RCM to the complete system life cycle is a resource intensive

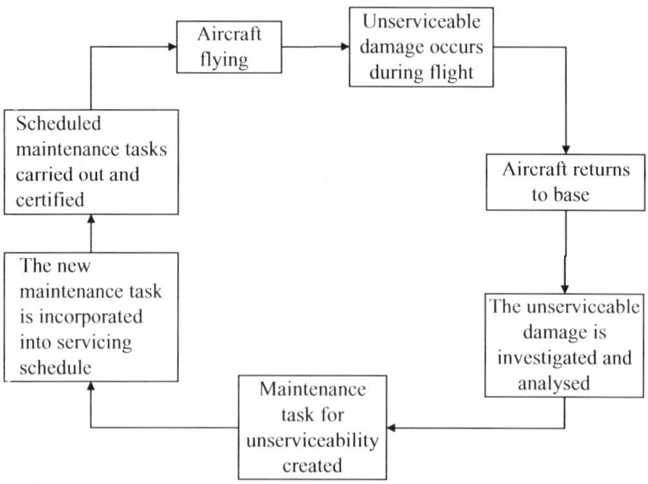

Figure 5.1-1 Typical work flow for in-service maintenance of aircraft

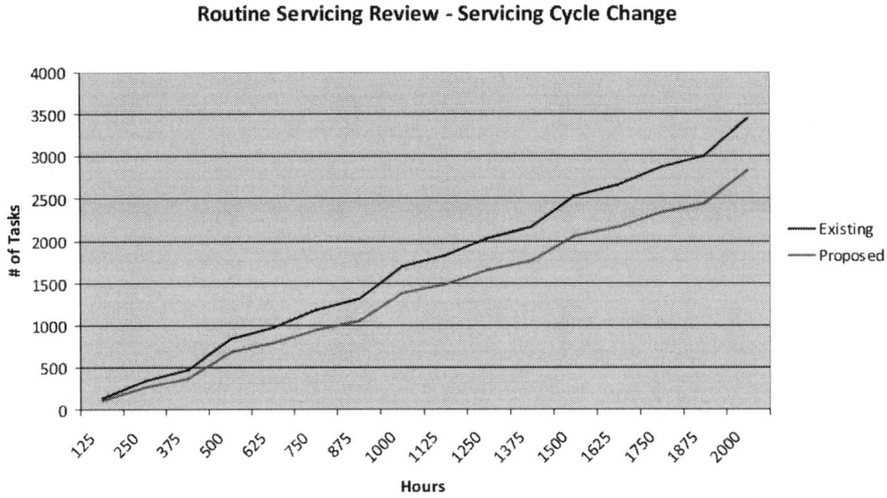

Figure 5.1-2 Application of RCM to lead-in fighter routine scheduling; 26% reduction in tasks

process and may not be cost effective in all cases. Therefore, the RCM analysis must be tailored to ensure the analysis goals are met in the most cost-effective manner.

When conducting in-service RCM, it is important to adequately identify the analysis requirements and prioritise the systems and items to be analysed. Some of the approaches used in applying RCM in-service include:

- *Stake-in-the-ground method.* This is a minimum-initial-effort method. It assumes most current PM tasks are reasonably justified and therefore, the application of RCM will be via proactive sustaining efforts.
- *High-profile analysis.* Similarly to the previous method, this consists of using proactive efforts such as analysing high-cost drivers or systems/items with low reliability characteristics.
- *Back-fill method.* This is a medium-level effort. It assumes that the current PM program adequately covers all potential failure modes and that there may be some PM being performed that may not be required. A list of items and/or functions to be analysed and validated is developed from the existing PM tasks.
- *Complete analysis.* This approach encompasses a total review of the application's systems, subsystems and associated items. It requires the highest degree of effort and should only be considered when potential returns are high (i.e. for programs with significant life remaining, high maintenance costs, or very low reliability).

5.2 In-service engineering support

The in-service engineering authority works with the stakeholders on numerous issues. Some related positions are:

- *Operator.* This is the equipment user and priority customer. The in-service engineering responsibility is to keep the product safe and effective for this customer's use. The accuracy of operator manuals is vital.

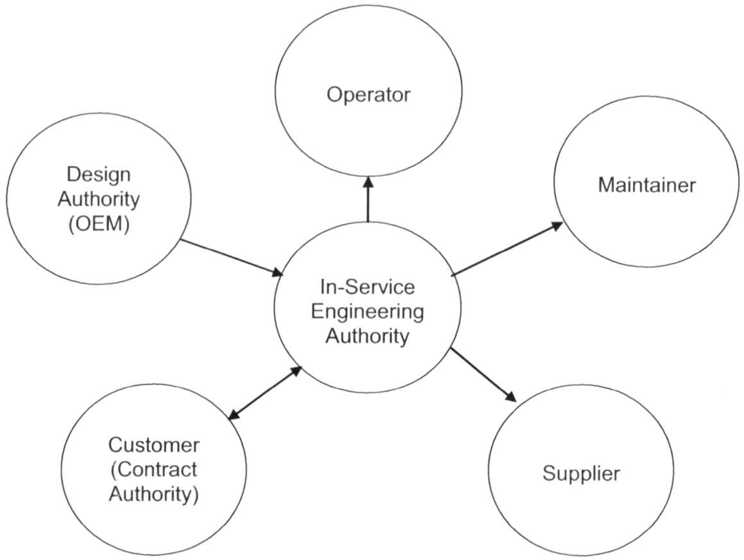

Figure 5.2-1 In-service engineering relationships with stakeholders

- *Maintainer.* This is the on-site system maintainer (e.g. on aircraft). The subsystem or equipment maintainer will be off-site (e.g. off-aircraft, performing bench repairs etc.). This customer is reliant on engineering advice to assure product serviceability (e.g. maintenance manuals, support equipment definition).
- *Supplier.* This is the party responsible for supplying parts defined by the product-configuration baseline. Parts unavailability will be a problem at many points in a product's life. Engineering solutions to logistics problems are often required.
- *Customer contract authority.* This is the party responsible for managing the in-service contractual requirements, typically flown over from acquisition. Manages latent defects under the contract and is responsible for ensuring contract adherence.
- *Product design authority.* This is the party initially responsible for designing the product. Not always the same organisation as the in-service engineering authority. Able to provide crucial information about equipment design, which means that strong communications links are extremely important.

These relationships can be illustrated diagrammatically as in Figure 5.2-1.

In-service engineering is part of the broader picture of product-logistics support. Customers have different needs, but the overall goal is the same: to support the product's use in its defined role.

5.3 In-service system safety

System safety is defined as the application of engineering and management principles, criteria and techniques to optimise all aspects of safety within the constraints of operational effectiveness, time and cost throughout all phases of the system life cycle (Browning and Honour, 2008) (Figure 5.3-1).

5.3.1 Categories of risks

As shown in Figure 5.3-1, the primary means of ensuring system safety is to assess risks. Individual risks, which are determined by assessment of severity in conjunction with the probability of a hazard, are expressed in terms of hazard-risk level and/or hazard risk index (HRI). Where there is correlation between an issue and events within the Hawk LIF Fault Trees, the fault trees will be used as the basis for identifying related hazard(s) and associated risk(s). This process may involve recalculation of the probability using the in-service failure rate for defined event(s). Where the severity and/or the probability of the hazard varies from the definition in the applicable fault tree(s), the risk level will be assessed using the risk-assessment matrices provided in Tables 5.3.1-1, 5.3.1-2 and 5.3.1-3. Similarly, where the issue and related events have no direct correlation with the fault trees, credible hazards will be identified. Severity, probability and risk level will be defined using the risk-assessment matrices provided below.

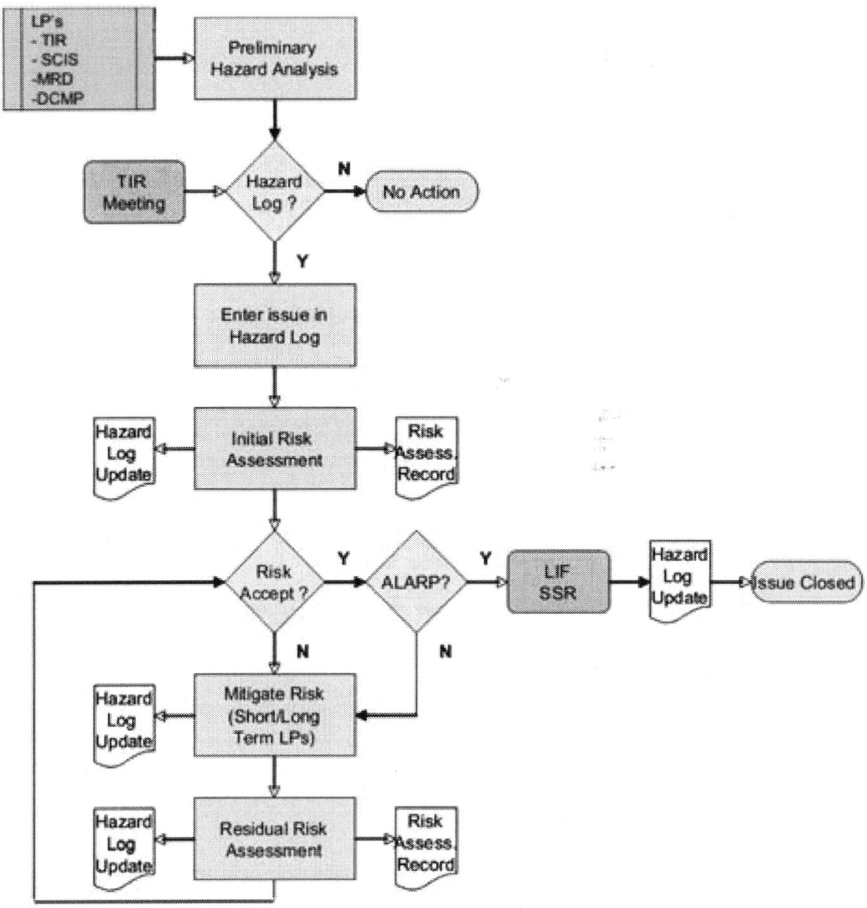

Figure 5.3-1 Example of risk-analysis process

Table 5.3.1-1 Hazard-severity categories

Description	Category	Consequences
Catastrophic	I	Death, aircraft/system loss, or severe environmental damage.
Critical	II	Severe injury, severe occupational illness, or major aircraft/ system or environmental damage.
Marginal	III	Minor injury, minor occupational illness, or minor aircraft/ system or environmental damage.
Negligible	IV	Less-than-minor injury, minor occupational illness, less-than-minor aircraft/system or environmental damage.

Table 5.3.1-2 Hazard probability categories

Description	Level	Specific item	Fleet or inventory	Failure probability per flight hour
Frequent	A	Likely to occur frequently.	Continuously experienced.	$\geq 1.0 \times 10^{-3}$
Probable	B	Will occur several times in the life of each item.	Will occur frequently.	$<1.0 \times 10^{-3}$ to $\geq 1.0 \times 10^{-4}$
Occasional	C	Likely to occur in life of a specific item.	Will occur several times.	$<1.0 \times 10^{-4}$ to $\geq 1.0 \times 10^{-5}$
Remote	D	Unlikely, but possible to occur in the life of an item.	Unlikely, but can be reasonably expected to occur.	$<1.0 \times 10^{-5}$ to $\geq 1.0 \times 10^{-6}$
Improbable	E	So unlikely, it can be assumed that the occurrence may not be experienced.	Unlikely to occur, but possible.	$<1.0 \times 10^{-6}$

The probability of a hazardous event occurring during the planned life of an aircraft is defined as potential occurrence per flight hour (averaged over the fleet life).

Table 5.3.1-3 Hazard-risk matrix

Probability category	Severity category			
	Catastrophic	Critical	Marginal	Negligible
Frequent	1	3	7	13
Probable	2	6	9	16
Occasional	4	6	11	18
Remote	8	10	14	19
Improbable	12	15	17	20

5.3.2 Initial hazard-risk assessment

Risk assessment will be conducted on initial identification of a hazard. Where applicable, they will take into account the in-service circumstances that may have led to recording the issue in the hazard log. Examples include the probability of hazard due to in-service failure rates and/or the actual consequences of the identified hazard(s). Where the initial assessment is conducted as part of a logistic process, reference to that assessment and the outcome (HRI) must be recorded in the hazard log. Where there is no existing risk assessment, LIFMO support-system engineering will conduct the assessment and record the analysis within a record of the entry or as a standalone record on file. Here again, reference to the assessment and the outcome (HRI) will be recorded in the hazard log. The management is to be immediately made aware of hazards where the initial assessment identifies a HRI of less than 10.

5.3.3 Hazard-risk mitigation

Mitigation actions are developed and implemented by the hazard actionee(s), primarily through existing LIFMO processes. The assessed HRI can be utilised to provide a basis for prioritising and defining the level of effort of the mitigation activities. In many cases, the mitigation actions will include short-term interim activities that bring the risk to an acceptable level, followed by longer-term activities that provide a permanent and ALARP solution. It should also be noted that if the initial risk assessment identifies that the risk level is acceptable and it is considered to be as low as reasonably practicable (ALARP), no mitigation is required. In this case the residual HRI recorded is the same as the initial HRI, and the hazard-log issue can be recommended for closure.

Further assessment of the hazard risk will be conducted to determine the effectiveness of mitigation actions, both short term and long term, in reducing the risk. The residual risk will be re-assessed and updated on completion of such mitigation actions. The residual risk level will be used as the basis for recommending whether the hazard log entry should be closed or remain open. The acceptance criteria defined in Table 5.3.1-3 are to be used in the determination.

5.3.4 Hazard-risk acceptance

Following each risk assessment, initial or residual, acceptance of the hazard-risk level is conducted and recorded. A statement of acceptance is to be recorded as applicable, using the criteria provided in Table 5.3.4-1. The following details are provided for additional guidance.

- Where the issue and associated events have clear correlation with the Hawk LIF Fault Trees, the resultant hazard-risk level is considered to be acceptable if it is within that defined by the fault-tree analysis and is accepted within the Hawk LIF Operational Safety Case.
- Where the risk level is outside that defined by the fault-tree analysis and the operational safety case, or there is no direct correlation with the fault-trees- analysis outcomes, action will be taken to eliminate identified hazards or reduce them to an acceptable hazard-risk level as defined in Table 5.3.4-1.

Table 5.3.4-1 **Hazard-risk resolution**

HRI	Risk level	Acceptance criteria
1–5 inc.	High	Unacceptable risk. Mandatory correction, elimination or control of hazard to an acceptable risk level.
6–9 inc.	Medium	Undesirable risk. Attempts should be made to eliminate or control to an acceptable risk level. Requires SDE/DAR acceptance if not corrected.
10–17 inc.	Moderate	Acceptable risk level by DE or above, with endorsement by the LIF System Safety Review.
18–20 inc.	Low	Acceptable risk level; review as required.

Although hazard-risk level may be assessed as acceptable, further mitigation actions to reduce the risk to ALARP may be warranted.

5.4 COTS-item obsolescence management

As significant issue in engineering change management is the increasing use of commercial-off-the-shelf (COTS) components in enduring assets such as defence systems, commercial aircraft, ships etc. COTS products have disadvantages and limitations, however. They are designed for the consumer market; that is, a relatively benign operating environment. Their reliability does not have to be very high given their short life cycles. Enduring assets, by contrast, are often subjected to extreme environments; hence their use is more economical where lives and valuables are at risk and a high level of reliability is required. The use of COTS products will require a careful trade-off between the benefits they offer and the limitations they may put on a system. Changes to COTS products, including patches and upgrades, incur costs that increase with the level of customisation required. These costs must be considered in a support plan to avoid support-cost overruns. The supported life of COTS products may be less than the typical life of enduring assets, which is often 20 years or more. A COTS product may become obsolete before the end of life of the system in which it is installed.

Obsolescence in this context refers to termination of production and support. Obsolescence of COTS products must be managed to minimise the operational risks. Supportability is a key area that requires special attention. Enduring systems that employ COTS technology will require a life-cycle support strategy that is different to what is traditionally used. The life cycle of a COTS product, which is typically 5 years or less, is out of sync with the life cycle of an enduring system, which is typically more than 20 years.

During the production phase of a COTS product, numerous minor modifications are introduced to correct problems identified with the product in the market or due to changes in component availability. These production configuration changes may result in incompatibilities between what would otherwise appear to be the same product.

Many COTS manufacturers do not apply configuration control. It is estimated that 70% of COTS technology will become unavailable on projects that require 20 years of support. Ministries of defence consider obsolescence as one of the highest risks in the support of enduring assets (Bil and Mo, 2013).

5.4.1 Obsolescence progression

Figure 5.4.1-1 shows the stages of COTS to obsolescence. The main COTS-risk trigger points are EOL, EOS, EOR and EOM, with different actions.

When a COTS product is nearing end of life (EOL; i.e. out of production) or end of service (EOS; i.e. no longer supported by the manufacturer), the effects on product supportability must be examined at both the product and system levels to determine what actions are needed. It is not a necessary immediately by newer versions of those products at EOL or EOS. The effects can range from no impact to high impact. The obsolescence-support options that are available to address these impacts can range from taking no action to making a major system redesign (Table 5.4.1-1).

It is obvious that at EOR and EOM, the capability of the enduring asset will diminish if actions have not been taken earlier.

5.4.2 Supportability of COTS products

Because COTS products are not specially designed as fit-for-the-purpose components of defence systems, many issues may exist with supportability. Questions to ask to determine whether such issues exist include:

- Are the data import/export interfaces open/standard and easy to use? This is important because if the system must be changed, at least the data contained in the COTS component can be archived.
- Are any control interfaces that will be used relatively standard? This question refers to custom commands or commands controlled by (for example) a well-known scripting language, which are less important than data interfaces.

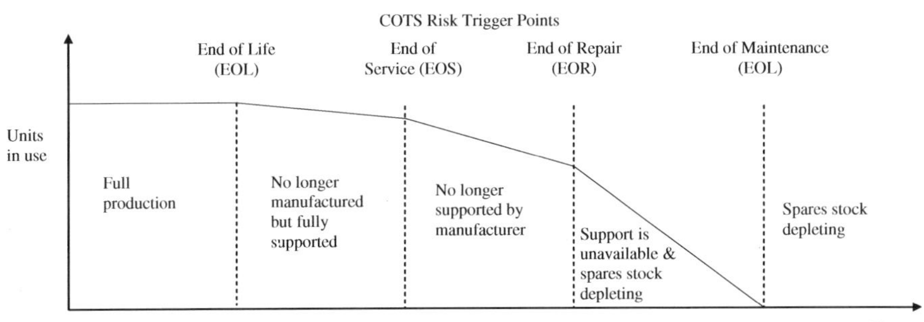

Figure 5.4.1-1 COTS obsolescence progression

Table 5.4.1-1 **Impacts of COTS-product obsolescence**

Impact	Actions
No impact	None. A COTS product is reliable and there are sufficient spares or repair parts on the market.
Low impact	The supportability status of a COTS product is projected to become unreliable and the product must eventually be replaced. Other products are available with no interface problems to other products in the system.
Medium impact	Similar to low impact, but with some serious compatibility issues. A major redesign will probably be necessary, including lifetime buy and other actions for the existing product.
High impact	As with low and medium impact, the supportability status of the COTS product is projected to become unacceptable and the product must eventually be replaced. There are no compatible replacement products or technologies available on the market. This situation typically calls for a major redesign.

- How standalone is the COTS-item application? If the COTS item's internal operating logic relies on specific operating-system calls, specific drivers or other underlying applications, the COTS items must be supported as a bundle.
- What support options does the supplying vendor offer? This consideration explores potential migration to another platform if the circumstance requires.
- Under what sort of frequency does the COTS product undergo updates? This interval will affect operation and maintenance costs.
- How does the vendor notify of upgrades and patches? Sometimes, these can be delayed until a major change is required. Doing so will save cost of ownership.
- How is the product activated or reactivated? The state at which the COTS product can be activated after power down or repair must be known.
- What is the desire in terms of updating? How does the validation of updating occur? The latter can be a time-consuming process.
- How is the link to a specific version of COTS product specified? If there are changes in other parts of the defence system but not the COTS product, a link to the required functionality is vital to the new defence system.
- What is the longevity of the vendor and of this product? The defence system lasts for many years but if the vendor disappears quickly due to business reasons, how can this COTS product be supported?

COTS item and technology insertion in enduring systems is not a simple plug-and-play scenario. Proper management strategy is required to define the support architecture of the product and the service systems that will ensure sustainability of operation.

5.4.3 Obsolescence strategy

A good obsolescence strategy has three dimensions that measure impact: when obsolescence occurs, the probability of obsolescence occurring, and the cost associated

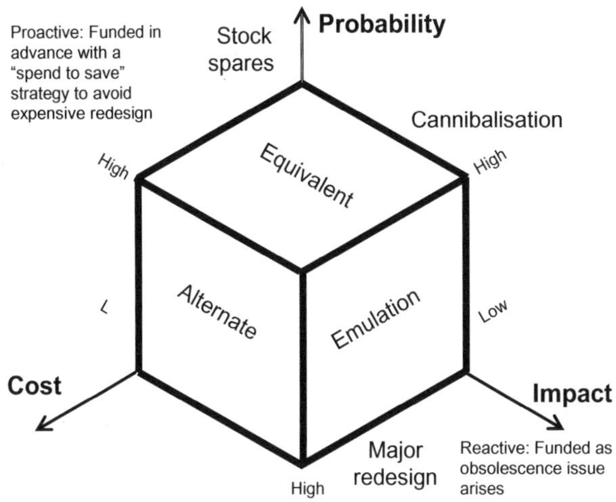

Figure 5.4.3-1 Dimensions of obsolescence

with the given strategies (Figure 5.4.3-1). The cost may be a combination of financial and non-financial factors. Strategies may include stock spares, cannibalisation, last-time buy (LTB), substantial equivalent, emulation, alternate, redesign etc.

Using the three dimensions, project directors or managers decide on a proactive or reactive approach for each dimension to minimise risk.

Mitigation	Approach
Proactive mitigation	Agreements with suppliers to ensure sub-component longevity, or considering obsolescence at the concept phase to mark key components for proactive management.
Reactive options	Making an LTB or dealing with any obsolescence issue reactively; possible ramifications include major redesigns or non-recurring engineering (NRE) costs.

A decision is made to take a proactive or reactive approach to obsolescence management of individual items based on the responses to a number of criteria that take into account the triplet of impact, cost and probability (Figure 5.4.3-2). A reactive response may be suitable if the level of classification is low for the three criteria; otherwise, a proactive response is most suitable. Figure 5.4.3-3 shows a decision logic for proactive and reactive approaches.

The obsolescence management plan includes:

- Identification of COTS-product obsolescence-treatment strategies (see above) that may be applicable to the system under consideration.
- Establishment of a COTS-product-selection actions checklist to ensure that obsolescence risks are considered early in the design phases.

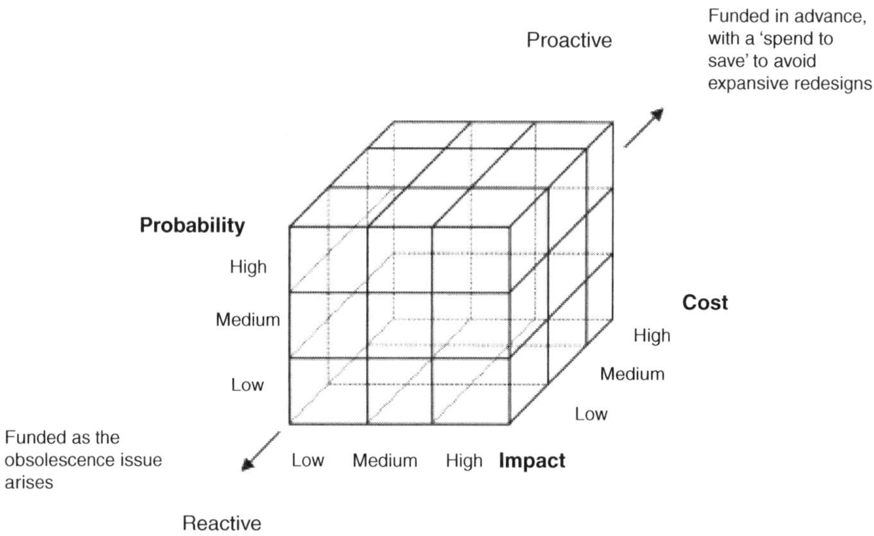

Figure 5.4.3-2 Risk-assessment diagram

- Performance of a qualitative analysis of the candidate COTS items to classify them into different categories to enable the selection of a cost-effective COTS-item obsolescence-treatment strategy.
- Performance of a quantitative analysis for spares requirements for all COTS items introduced into the system to establish the basis for assessing life-of-type (LOT) spares requirements (regardless of the length of the selected LOT).

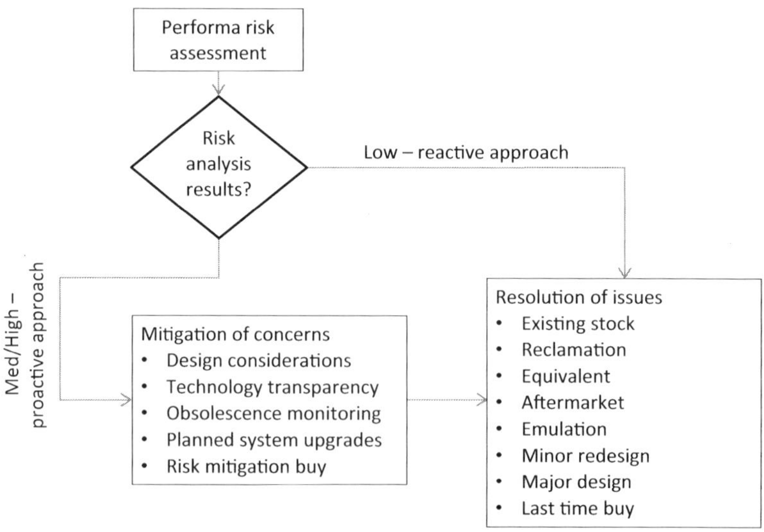

Figure 5.4.3-3 Reactive versus proactive approach

References

Bil, C., Mo, J.P.T., 2013. Obsolescence management of commercial-off-the-shelf (COTS) in defence systems. In: Stjepandić, J., Rock, G., Bil, C. (Eds.), Concurrent Engineering Approaches for Sustainable Product Development in a Multi-Disciplinary Environment. XLVI, 1217, pp. 621–632.

Browning, T.R., Honour, E.C., 2008. Measuring the life-cycle value of enduring systems. Sys. Eng. 11 (3), 187–202.

Johnstone, S., Dainty, A., Wilkinson, A., 2009. Integrating products and services through life: an aerospace experience. Int. J. Oper. Prod. Manage. 29 (5), 520–538.

Systems prototyping and testing 6

Chapter Outline

Support system design cannot be tested easily. The 'laboratory' is often the real application scenario. Some researchers use case-study research methodology, but their ability to do so depends on the existence of comparable cases from which principles can be detected and drawn. This chapter uses an example to illustrate how a support system can be prototyped and tested as a pilot study in an actual industrial environment.

6.1 Prototyping

This illustrative example is a business-case study of the application of radio-frequency identification (RFID) technology to track goods through a supply chain (Sarma et al., 2001). The aim is to economically provide greater security and product authentication from point of origin. Due to great demand, many individual suppliers of RFID systems have been selling their proprietary solutions in the market. Unfortunately, these systems do not work interchangeably in supply chains with several proprietary systems and hence are unable to generate real benefits to the community.

EPC networks are special RFID implementations that are suitable for fast-moving consumer-goods (FMCG) tracking (CSIRO, 2006). However, Australian companies are generally small such that the efficiency savings within a single enterprise are not sufficient to substantiate a viable business case. In order for Australia to secure real benefit from the development, a community-based approach that enables a large number of organisations and companies in all kinds of supply chains to be EPC-network-ready is required. At this time, there is not enough information to help companies understand how the technology should be used in their environments (including supply chains) and to assess what benefits (tangible and intangible) it can bring as well as the companies' readiness to implement it. Thus, an experiment to capture the knowledge that would help these companies to apply the technology appropriately and to assess the benefits of doing so in an inter-enterprise operating environment is necessary.

A model of how EPC operates among independent organisations and the EPC network processes will provide Australian businesses with insights into this advanced B2B e-commerce technology.

Engineering Systems Acquisition and Support

6.2 Experimental design

The support system involves a large number of industry players across multiple segments. Implementations of RFID technology so far have concentrated only on the physical layer (i.e. tags and readers), but the requirements of the information flow for supporting-business needs in inter-company transactions are not addressed. This experiment aims to demonstrate a complete support system of tracking and tracing goods as they move through the FMCG supply chain. Although the involvement of multiple organisations (and multiple links between multiple organisations) introduces an additional level of complexity, it also gives the implementation credibility that other trials have lacked because they did not capture representative behaviours across a whole supply chain and industry.

In the preparation of the pilot, the following question arises: 'How many of the EPC items do we need to process and over what time period, to ensure a valid demonstration of the EPC network'? The number of items may be expressed as the number of tags or the number of events in each scenario (e.g. a pallet of tagged cases is one event).

This question has implications about the length (duration) of the trial. Unfortunately, the answer is not straightforward because it depends on how the experiments are designed (i.e. the nature of the statistical item) and the process of validation.

We attempt to use an example to illustrate how the concept of experimental validation is applied and how the parameters can be determined.

At the dispatch station of P&G, boxes of pentene are EPC tagged and put on an EPC preregistered pallet. The loaded pallet is pushed through an EPC reading gate where the EPC tags on the boxes and the pallet are read. The EPC system has a successful pass if all EPC-tagged boxes and the pallet are read in a single, unrepeated pass. If the expected read rate is that in 100 passes only 1 pass fails, we need to test the hypothesis as follows:

H0: The system has a failure rate of $p\%$ (i.e. p failed passes per 100 passes).

H1: The system has a failure rate greater than $p\%$.

It is assumed that $p = 1$. The hypothesis is tested with 95% confidence level.

Assuming the prototype system is tested with 100 runs, the number of failures will be given by:

$$\bar{x} = np = 100 \times 1\% = 1$$

The variance of the sampling distribution

$$V = npq = 100 \times 1\% \times (1 - 1\%) = 0.99$$
$$\sigma = \sqrt{V} = \sqrt{0.99} = 0.99499$$

The confidence limits for the hypothesis are then:

$$\text{UCL} = \bar{x} + z_{\alpha/2}\,\sigma = 1 + 1.95996 \text{ X } 0.99499 = 2.9501$$
$$\text{LCL} = \bar{x} - z_{\alpha/2}\,\sigma = 0.9501$$

The earlier calculations mean that with a confidence level of 95%, the test is expected to show the read-error rate ranges from 0.9501% to 2.95091%. In other words, we are confident that the success system read rate is between 99.05% and 97.05%. If we have a success-read rate of 95%, this means the system does not meet our requirement and H0 should be rejected in favour of H1. However, if the test shows a success-system read rate of 98%, then we can accept H0. The decision on the success-read rate is based on commercial risks.

The confidence limits indicate the maximum error of estimation for the real-read error. The interpretation of this maximum error ε_{max} is that we are $1 - \alpha$ sure that the population mean of the system-read rate falls within the interval on either side of the estimated mean by the maximum error. Of course, in any sampling distribution, it is still possible that the error ε can be larger than the maximum error ε_{max} when the interval is fixed. In other words, there is always a risk associated with an interval estimate. Because the maximum error is related to the sample size n, the previous interpretation of the risk gives a method to determine the required sample size from a given level of confidence and an allowable error.

Using this scenario, if we do not want the error of estimation ε_{max} to exceed ±1% (i.e. if we want a 98–100% success rate), the number of samples required is:

$$n = (z_{\alpha/2}\,\sigma / \varepsilon_{max})^2 = (1.959962 \times 0.99499 / 0.01)^2 = 195.01$$

We have only 100 passes, which is not sufficient to sustain the confidence level we want to achieve. We need to do 95.01 (i.e. 96) more passes. After 196 passes, if the failure rate is more than $\sigma = \sqrt{V}$ within 1% (i.e. 3 failures within 385 tests; we can't have 3.8 failures so the number must be rounded down), our results are not conclusive and the sample size must be recalculated.

The time required to do the test depends on how frequently the passes are made. The mathematics is simple: If we have 10 pallets passing through the gate per hour, then we need 19.501 hours, or about 2.5 working days.

The general rule of experimental validation is that the events are random and independent. The passes in the previous scenario should therefore be managed independently. By this we mean that no two events should be linked by a corrective action. For example, if a pass fails and someone pushes the pallet back to go through the gate again (without restacking the pallet in a different order of boxes), and even the second read is successful, we cannot use this result. We still need to count the read as fail and discard the second read.

6.3 Physical testing

The National EPC Network Demonstrator Project (NDP) aimed to identify the business benefits of sharing information securely using the Electronic Product Code (EPC) Network. The shared information was meant to provide authentication to interacting parties and to enhance their ability to track and trace the movements of goods within the entire supply chain; these movements involved transactions among multiple enterprises.

In order to capture the test data, the NDP website was set up on a global server and programmed to show how product information, containment (content), and history (track and trace) could be accessed using 'EPC' as the search keyword (Figure 6.3-1).

Because capturing RFID physical data is a critical activity in the physical system, hardware configuration was one of the major developments in the trial. To ensure that the RFID readers could capture any tag passing through the dock door of the distribution centres, the antenna configuration in NDP was a portal configuration with two antennas mounted on steel posts on each side of the door (Figure 6.3-2).

The 2 × 2 portal provides a uniform detecting field in the door-opening area. However, for some warehouses, tagged goods may stick to one side of the goods. In order to improve the readability of the system in such circumstances, an alternative configuration is developed by the consortium to concentrate all available antennae on the readers to a one-sided configuration (Figure 6.3-3).

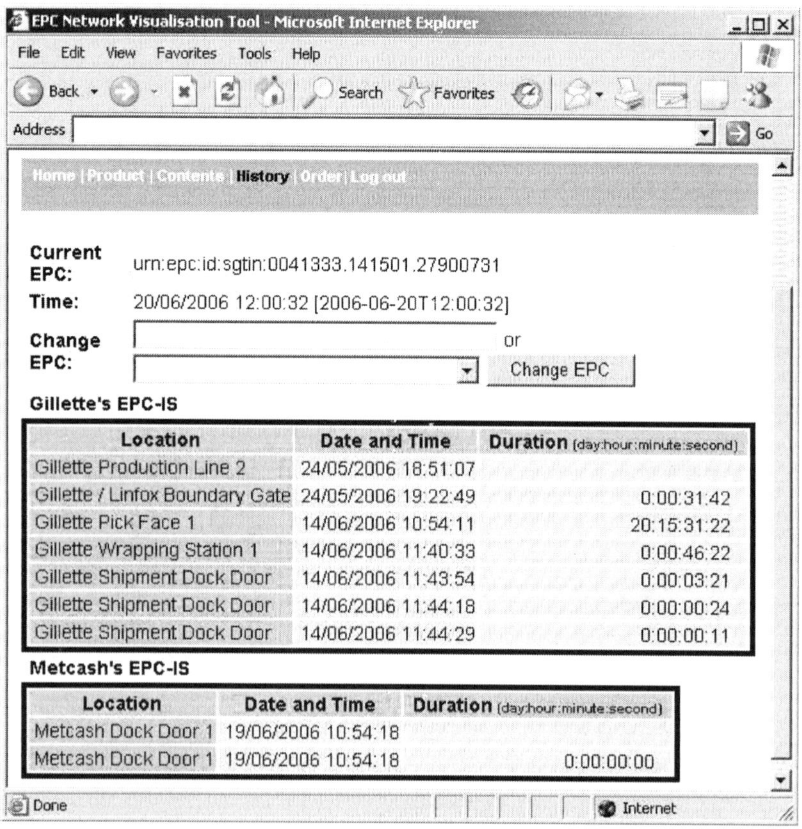

Figure 6.3-1 Track-and-trace for item urn:epc:id:sgtin:0041333.141501.27900731

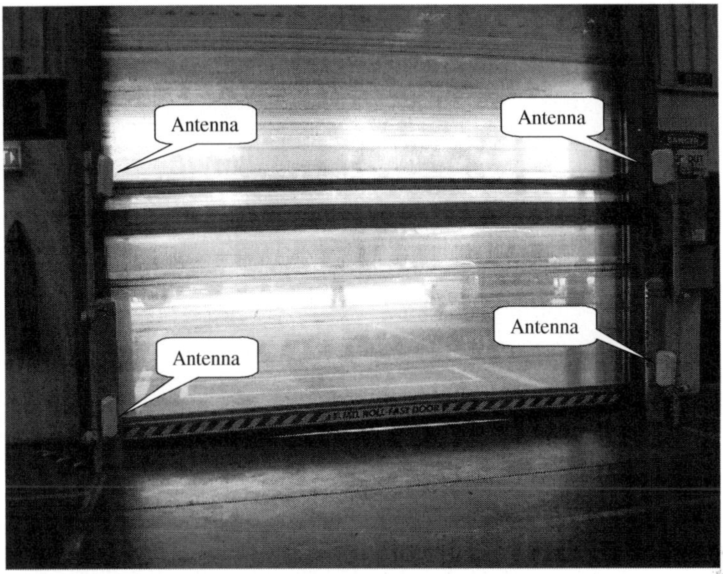

Figure 6.3-2 Dock-door portal configuration with 4 antennae

This example illustrates that support-system design varies greatly. Each hardware configuration must be designed specifically for a particular scenario (Mo et al., 2009). Because the physical setup-design process is still based on an ad hoc, time-consuming, trial-and-error approach, a prototyping test procedure is necessary to maximise the probability of success.

6.4 Virtual testing

If physical prototyping and testing are not feasible, virtual testing by simulation can be considered. The NDP has been tested using a simulated system (Kelton et al., 2007). Creation of the NDP simulation model starts with the Create module (here called *Metcash Inventory Control*). The first module in the submodel is called *Metcash Product Inventory Control*. When this entity is created, Metcash will check the inventories of all products. If the inventory exceeds a certain level, a PO won't be created and the entity will go into Dispose modules. This is one kind of whole-life flow that extends from Create to Dispose.

The establishment of NDP simulation model is based on the basic work flow and the information flow in Figure 6.4-1. At the beginning of each day, Metcash DC checks inventory (denoted here as $I(t)$, where t is time) to decide whether to place an order. If the inventory level is less than a constant s, Metcash DC orders up to another constant S. What this means is that Metcash DC places a PO that, if it were to arrive instantly, would raise the inventory level to exactly S.

Figure 6.3-3 Antennae mounted as one-sided configuration in NDP extension

Metcash's PO initiates most of activities at P&G, Gillette CHEP and Visy. Only Gillette Manufacturing is independent from Metcash's PO. However, Gillette Manufacturing's activities are based on the sales forecast and the production quantity will be based upon the PO quantity rather than upon the inventory quantity.

The NDP simulation model is further separated into five submodels. Each submodel represents a consortium member and has several functions for different types of work. All submodels and functions are listed in Table 6.4-1.

Before the simulation is set to run, some important parameters must be set up. The simulation time is 360 days. Initial values of variables should be given before the simulation is run, because they describe the situation at the time that the simulation is run. We assume that all empty-pallet and flat-carton request situations and PO situations are *0*. This means there is no request or PO at the start, irrespective of whether the inventories are lower than level *s*. All the initial values of inventories are listed in Table 6.4-2.

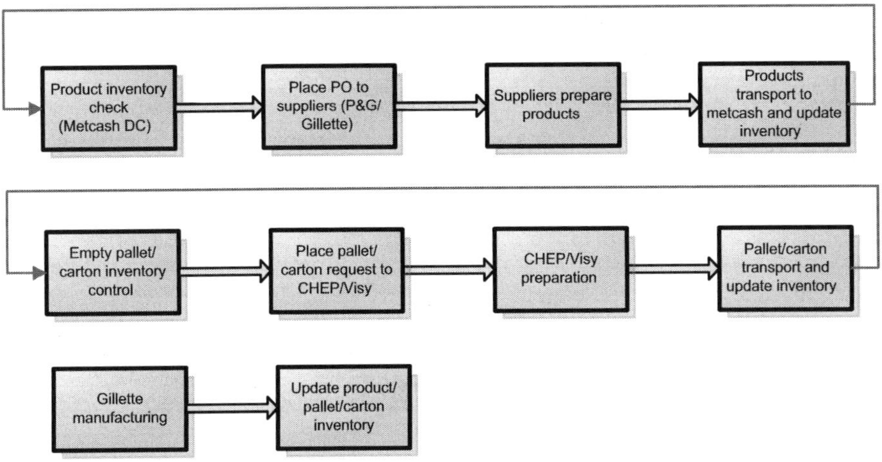

Figure 6.4-1 Basic work flow in NDP-simulation model

Table 6.4-1 NDP simulation modules

Module	Name	Brief description
Metcash DC	Metcash Receiving Process	To receive products from P&G and Gillette, and transport them to designated areas.
	Metcash Product Inventory Control	To monitor and take actions that control the inventory level at Metcash.
	Metcash Dispatch Process	NDP Simulation Model doesn't keep trace the pallets and cartons, so the submodel just describes the customer arriving with some demand attributes, and updates the inventories after the products are dispatched.
Gillette	Gillette Manufacturing	There are two series of products in the Gillette plant. They are produced as small packs, but entities in the simulation model are expressed as pallet units. Every Process module relates Use cases in EPC Infrastructures and Implementation Processes. [10]
	Gillette Pack and Wrap Pallets for Customer Order	This is a complicated function in the simulation model because it includes two important activities: packing and wrapping, merged as one logic control.
	Gillette Flat Cartons receiving and Empty Pallets Returning Process & Gillette Empty Pallets Receiving Process	This is a simple function: flat cartons or empty pallets arrive at Gillette. Empty pallets and flat cartons are transferred as stacks and pallets, respectively.
	Gillette Pallet and Carton Request	This is a logic control for Gillette requesting empty pallets and flat cartons. It is similar to Metcash Product Inventory Control.

(Continued)

Table 6.4-1 NDP simulation modules *(cont.)*

Module	Name	Brief description
P&G	P&G Pallets Receiving Process	Pallets will be unloaded, and transferred through reader gantry; inventory will be updated automatically and the pallet-request situation will be changed.
	P&G Use Cases	This function strictly follows EPC Infrastructures and Implementation Processes. Related use cases are listed in Appendix. PO will be issued by administrators.
	P&G Pallet Request	This is another logic control of the (*s*, *S*) inventory model for P&G placing request for empty pallets (similar to the Gillette pallet request).
CHEP	CHEP Use Cases	There is only one function in CHEP, which is to prepare empty pallets and transport them to customers.
Visy	Visy Use Cases	There is one logic control of the (*s*, *S*) inventory model for Visy placing request of empty pallets. Visy checks empty-pallet inventory once a day.

Processing Time means the time interval that begins when a PO is placed and ends when the required products are received. It includes delays at a value-added process, wait time in a queue and transfer time. The application of RFID will increase time of some processes, such as placing tags on cartons, pushing cartons through gravity feedline, scanning tags on pick faces etc. The application of RFID will also decrease the time of some other process, such as administration work after product and pallet arrivals, document checks etc.

The cost of labour varies slightly at different locations in the country. The actual labour cost per hour is used as shown in Table 6.4-3.

The main measure we will use to assess the authenticity of the simulation is examination of the three pilot cases: P&G Product Inventory, Gillette AA4 Inventory and

Table 6.4-2 The initial values of variables

Variable name	Initial value
P&G Empty-Pallet Inventory	30
Gillette Empty-Pallet Inventory	100
Gillette Empty-Carton Inventory	40
Visy Empty-Pallet Inventory	50
P&G Product Inventory	30
Gillette Mach3 Inventory	15
Gillette AA4 Inventory	15

Table 6.4-3 Labour cost list

Company	Labour cost/hour ($)
CHEP	32
P&G	28
Gillette	25
Visy	23
Metcash	31

Gillette Mach3 Inventory. The inventories rebound between proposed levels and there are rare shortages or overstocks. The plots are shown in Figure 6.4-2a–c.

We calculate PO Processing Time in days by the function

PO Processing Time = VA Time + NVA Time + Wait Time + Transfer Time

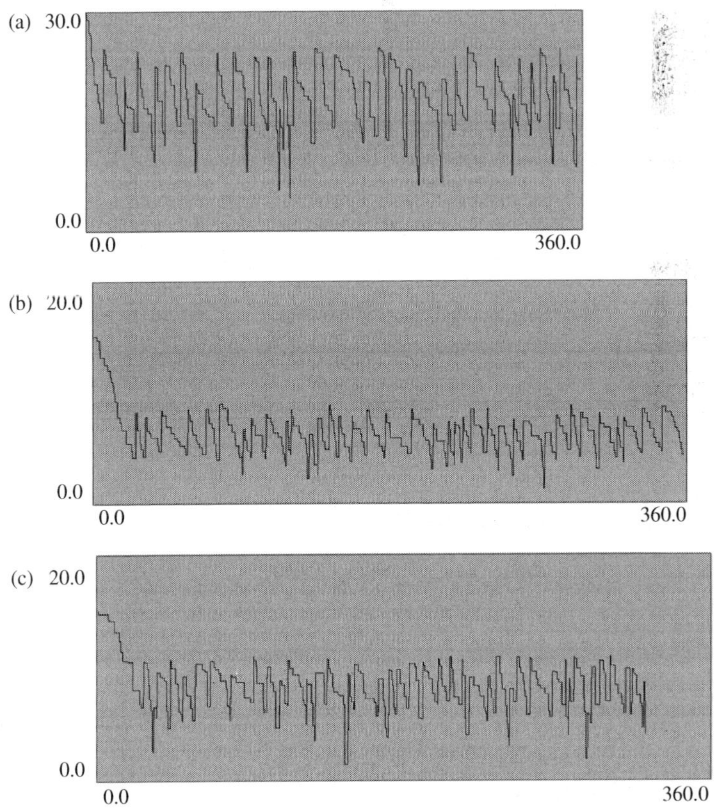

Figure 6.4-2 (a) The P&G product inventory plot, (b) the Gillette AA4 inventory plot and (c) the Gillette Mach3 inventory plot

Table 6.4-4 PO processing time (days)

	Average PO processing time (days)	Accumulated total PO processing time (days)	Average VA time (days)	Accumulated VA time (days)	Average wait time (days)	Accumulated wait time (days)
AA4 PO	1.244	242.619	0.0685	13.365	0.157	30.6345
Mach3 PO	1.241	323.901	0.0535	13.973	0.147	38.2365
PG PO	1.798	61.118	0.5579	18.969	0.000049	0.001666

where

- VA Time means value-added time. It is accumulated when an entity incurs a delay at a value-added process.
- NVA Time means non-value-added time. It is accumulated when the entity incurs a delay at a non-value-added process.
- Wait time is accumulated when the entity incurs a delay at a process whose allocation has been designated as wait, or when the entity resides in a queue until it exits the queue.
- Transfer Time is accumulated when the entity incurs a delay at a process whose allocation has been designated as transfer. By default, all time spent using material handling from the Advanced Transfer panel (e.g. a conveyor or transporter) is specified as Transfer Time.

The Average Purchase Order Processing Time, VA time and Wait Time are then computed from the simulation as shown in Table 6.4-4.

The labour cost can then be evaluated by the function:

$$\text{Labor Cost} = \sum \text{Average Number Busy} \times 360 \times \text{Daily Pay}$$

Table 6.4-5 shows the resource implication.

Table 6.4-5 Total labour cost

	Average number busy (days)	Total busy time (days)	Daily pay ($)	Sum ($)
CHEP Admin	0.00688	2.47764	256	634
CHEP Labour	0.00796	2.86654	256	734
Gillette Admin	0.00556	2.00012	200	400
Gillette Labour	2.5669	924.084	200	1848
Metcash Labour	0.00369	1.32761	248	329
PG Admin	0.00179	0.64445	224	144
PG Labour	0.04729	17.0248	224	3814
Visy Admin	0.00471	1.69695	184	312
Visy Labour	0.01471	5.29593	184	974
Total Cost	$192,158			

The savings of $192,158 is the overall labour cost savings if the NDP continues for a year. Other costs such as overhead and materials costs are not included. Therefore, this figure represents only a tiny section of the potential savings of the supply chain. Given that there are thousands of products handled by these 3 consortium companies, the savings will be significant if the EPC technology is rolled out to the whole supply chain.

References

CSIRO, 2006. EPC Network TM Australia Demonstrator Project Report. Impetus 2006, 25–26 July, Melbourne, GS1 Australia, CSIRO. Soft copy of the publication is available on request from http://www.gs1au.org/.

Kelton, W.D., Sadowski, R.P., Sturrock, D.T., 2007. Simulation with Arena, fourth ed. McGraw-Hill, ISBN 9780071286046.

Mo, J.P.T., Gajzer, S., Fane, M., Wind, G., Snioch, T., Larnach, K., et al., 2009. Process integration for paperless delivery using EPC compliance technology. J. Manuf. Tech. Manage. 20 (6), 866–886.

Sarma, S., Brock, D., Engels, D., 2001. Radio frequency identification and the electronic product code. IEEE Micro Nov–Dec, 50–54.

Operations support and logistics 7

Chapter Outline

The availability of effective global communication facilities in the twenty-first century has changed the goals of many businesses. They need to remain competitive by extending products and services to satisfy individual customers' requirements, completely packaged and operated globally. Networks of enterprises are formed to operate across time and space with worldwide distributed functions such as customer support, logistics, engineering, quality assurance, supply-chain management and so on.

Vast amounts of information can be generated by computers nowadays but these data cannot be useful unless they are organised and retained in the company. Rapid changes in enterprises for operations support and logistics make it difficult for companies to maintain a consistent knowledge base for supporting their businesses. Part of these fundamental changes in enterprises involves determining their operating characteristics in ways that provide methodologies on the modelling and analysis of their global operations.

The use of models to represent processes and activities in enterprises will help communication among different levels of operations and enhance cooperation among various parts of the operations, particularly their support and logistics systems. Models are also used for initial design of systems to ensure that the best possible performance can be obtained at the shortest possible time. The requirements of a working model are:

- Ability to represent systems both qualitatively and quantitatively.
- Possibility of analysis to produce results for decision (formalism).

Engineering Systems Acquisition and Support

- Ease of implementation; technologies to realise systems may include hardware and micro-programmed software.
- Knowledge sharing that is graphical, for ease of communication, to ensure that information about the system can be documented and understood by others.
- Ability to be managed by a project team, simplicity of construction.
- Ability to analyse in terms of time domain; inclusion of discrete events and dynamics.

7.1 Process modelling

Process modelling is the method of describing the steps of operating a system or carrying out an activity. Process models are not only required for describing the steps in a process for stakeholders to understand what activities are involved but are also used for planning, automation development and task assessment. The process model consists of block descriptions that lay out the essential steps of determining the operating characteristics, materials properties, performance attributes and other process characteristics such as time constraints. There are many process modelling methodologies, all of which are initiated through a theoretical framework that establishes their formalism. A formal modelling approach provides a structured means of achieving these goals and simplifies the complexity of communication. Over the years, many methodologies have been automated by computer-assisted modelling systems.

The goals of a process model should be:

- Descriptive
  - Track what actually happens during a process.
  - Take the point of view of an external observer who looks at the way a process has been performed and determines the improvements that must be made to make it perform more effectively or efficiently.
- Prescriptive
  - Define the desired processes and how they should/could/might be performed.
  - Establish rules, guidelines, and behaviour patterns which, if followed, would lead to the desired process performance. They can range from strict enforcement to flexible guidance.
- Explanatory
  - Provide explanations about the rationale of processes.
  - Explore and evaluate the several possible courses of action, based on rational arguments.
  - Establish an explicit link between processes and the requirements that the model must fulfil.
  - Predefine points at which data can be extracted for reporting purposes.

7.1.1 IDEF3

IDEF3 is one of the methods developed specifically for describing processes. IDEF stands for ICAM DEFinition methods (Mayer et al., 1995). It results from a series of studies carried out by the US Air Force in the 1970s. IDEF3 is loosely based upon *situation theory*. According to situation theory, the world contains several kinds of things: objects, situations, activities, processes and process activations. Objects, or individuals, are simply things (e.g. numbers, NC machines etc.) that are not situations (i.e. concrete or abstract objects that have properties and stand in relation to other objects *in* situations).

A situation is a real-world *event* across some portion of space and time. An example of a typical situation would be a meeting attended by a manager between 08:00 and 08:30 in Melbourne yesterday. This situation comprises some set of objects (the attendees) that have properties and stand in certain relations to each other in the situation. The most fundamental kind of information in situation theory is known as a *basic infon*. Such infons are of the form 'object a has property p', which is written (p a) in IDEF3's underlying 'elaboration language'; and 'objects a_1 ..., a_n stand in the relation r', which is written (r a_1 ..., a_n). Boolean and quantified combinations of infons are built up from infons.

In most physical systems one observes multiple occurrences of situations that are similar in some respect. In such cases, the similar situations are said to be of the same *type*. Situation types are thus general, repeatable patterns that can be exhibited by many different specific situations. The notion of an *activity* in IDEF3 is identified with an instance of a situation type.

It has been shown that there are many kinds of knowledge required in customer support. For example, in the case of supporting the diagnosis of a communication link, two kinds of knowledge are required. The first kind of knowledge consists of a description of a specific, well-defined task: installing a loopback tester, running a program, cleaning a cable end etc. The second kind of knowledge is represented by the temporal and logical *pattern* that these tasks must exhibit in order for them to collectively achieve the given end, viz., a successful diagnosis.

In process modelling, *language* can capture and distinguish both types of knowledge. The knowledge of individual tasks is linguistically captured as the content of particular activity boxes in the model of the dialogue process with the customer, and the temporal and logical pattern is captured in the junctions and precedence relations ('arrows') of the model. Such knowledge, of course, is as critical to a successful run of the procedure as the knowledge of the individual tasks.

The importance of types in the context of process modelling is that the semantic content (i.e. the *meaning*) of any process model involves types almost exclusively. A typical process is best thought of as a structured collection of activities that are related to one another by the temporal constraints in order to determine how the process can be instantiated. To represent processes intuitively, IDEF3 introduces several basic graphical symbols. For our purposes herein, to illustrate the meanings of the symbols in IDEF3 schematics we will use a simple procedural constraint between two activities in a process (Figure 7.1.1-1).

The activity instances (i.e. the situations) in an activation must occur within a single, finite interval of time that begins when the first instance in the activation begins and ends when the last instance ends. To determine whether a given collection of situations is an activation of a given process, it is useful to plot the general pattern of their occurrence over such an interval (Figure 7.1.1-2).

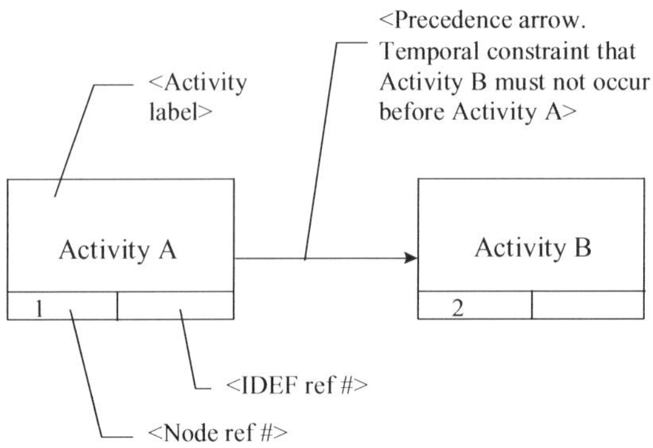

Figure 7.1.1-1 Basic precedence syntax in IDEF3

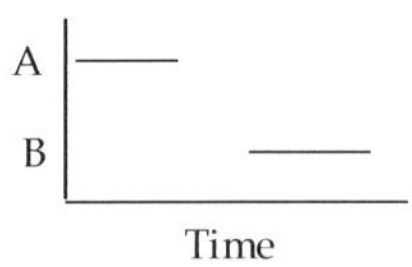

Figure 7.1.1-2 An activation plot for the process in Figure 7.1.1-1

IDEF3 diagrams (or schemata) are type-level descriptions of complex processes. Such processes are rarely linear. Typically, processes involve any or all of four general sorts of *branch points*:

1. Points at which a process diverges into multiple *parallel* subprocesses.
2. Points at which a process diverges into multiple (possibly nonexclusive) *alternative* subprocesses.
3. Points at which multiple *parallel* subprocesses converge into a single 'thread'.
4. Points at which multiple *alternative* subprocesses in the process converge into a single thread.

There are four general types of *junctions* to express the four general sorts of branch points. The first two sorts are expressed by *fan-out* junctions: *conjunctive* fan-out junctions represent points of divergence involving multiple parallel subprocesses, whereas *disjunctive* fan-out junctions represent points of divergence involving multiple alternative subprocesses. The last two sorts of branch points are expressed by *fan-in* junctions, which are of two types: *conjunctive* fan-in junctions represent points of convergence involving multiple parallel subprocesses, whereas *disjunctive* fan-in junctions represent points of convergence involving multiple alternative subprocesses. Conjunctive junctions are AND type, indicated by the symbol &. Disjunctive junctions may be either OR (denoted by the letter O) and XOR (denoted by the letter X).

To illustrate the modelling process of IDEF3, Figure 7.1.1-3 shows a model of a simple test process for the communication port of a CNC machine. In this process, after a loopback test one either simply removes the loopback tester and quits because the test has succeeded (e.g. it has shown that a suggested repair of the system has

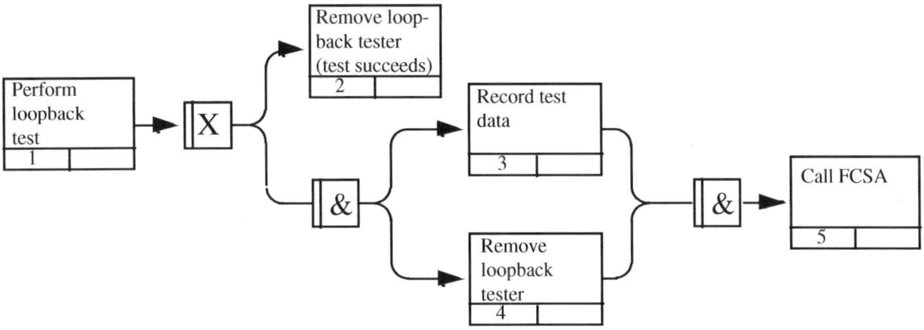

Figure 7.1.1-3 Simple loopback test process

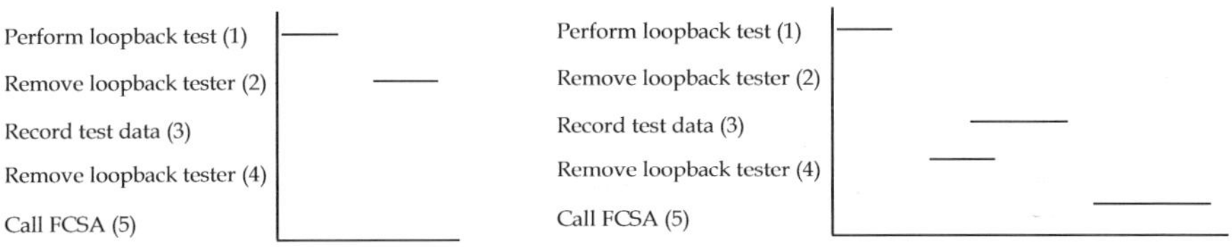

Figure 7.1.1-4 Valid activation plots for loopback test process

corrected an earlier problem) or one records the test data, removes the loopback tester and then calls FCSA for further assistance. An activation plot is the plot of a possible activation of this process if its structure matches that of the process. Note that there are therefore a number of activation plots corresponding to possible activations of this process, as illustrated in Figures 7.1.1-3 and 7.1.1-4.

A process often involves multiple occurrences of the same task or activity. However, the *position* of a task in the process is crucial for understanding what to do next. For example, one activity in the communications diagnostic process involves the running of a loopback test program. This activity is performed in the diagnostic procedure whenever the loopback is moved from one point to another. Consequently, which particular task follows a loopback test in a given instance of the communications diagnostic process depends on the process context in which the particular loopback test is performed. Thus, information about what to do *next* in the process cannot be contained within any rule or description of the loopback test proper – unless the description is specialised so as to distinguish between the different appearances of the task in the process. In IDEF3, the structure of the process model precisely captures such a distinction between the task and its context while at the same time allowing the re-use of the same task definition. This approach simplifies the application module development through reusing task definitions as components.

7.1.2 Petri net

Petri net is a theory of systems organisation originated by Carl Adam Petri in 1962. It has a strong emphasis on mathematical analysis that is formulated on the basis of set theory. Petri Net is a network model that has been used extensively by system designers and analysts to identify key processes within discrete event-processing environments (Aized, 2010). It consists of two key elements:

- *Net structure*: a weighted-bipartite directed graph that represents the static part of the systems.
- *Marking*: representation of a distributed overall state on the structure.

Formally, a Petri net can be defined as $N = (P, T, I, O, M)$
where

P = a finite set of places
T = a finite set of transitions
I = an input function such that $I : P \times T \rightarrow N$
O = an output function such that $O : T \times P \rightarrow N$
M = a marking vector with the same dimension as P, indicating the number of tokens at places in the network

Places are used to represent availability of resources, operation processes or conditions. They are pictured as circles. Transitions are used to model the events, start or termination of operations. They are pictured as bars.

To model a discrete-event-dynamic system, we need to take into account its states and the events leading to the state evolutions. In Petri nets, the state is described by means of a set of state variables representing local conditions. Hence, states are associated with places and transitions that are the entities of effect changes. Places and transitions are related through a weighted flow relation. The flow relation is modelled by arrows joining related places and transitions. A weight known as a 'token' can be assigned to the arc to indicate a numerical importance for that relationship.

We illustrate Petri net by an example. Consider the maintenance process of an aircraft at the hangar. To start the process, an aircraft is scheduled to be available for maintenance. When the hangar is vacant, the aircraft is moved into it. At the hangar, service and maintenance operations are performed upon the aircraft. A signal is passed to the hangar operator when all maintenance works are complete. The aircraft is then moved out of the hangar and is available for use. The hangar returns to the state of being available. This process is modelled by four places and two transitions in Figure 7.1.2-1.

This example assumes that the hangar can house one aircraft only. The maintenance process in a hangar that is large enough for two aircraft is modelled in Figure 7.1.2-2. The firing rule for t_1 is thus 'one token is taken from p_1 and p_4 respectively'. On the same basis, the firing rule for t_2 is 'distribute one token to p_3 and one token to p_4 after completion of service'.

Once the Petri net is created, reasoning at structural level can be applied to provide quick conclusions on the behaviour of the modelled system. Behavioural-level reasonings are computationally complex but provide accurate information for system design and improvement.

Complicated Petri nets can have hundreds of places and transitions. These nets can be structured in hierarchical format so that the process can be understood more readily.

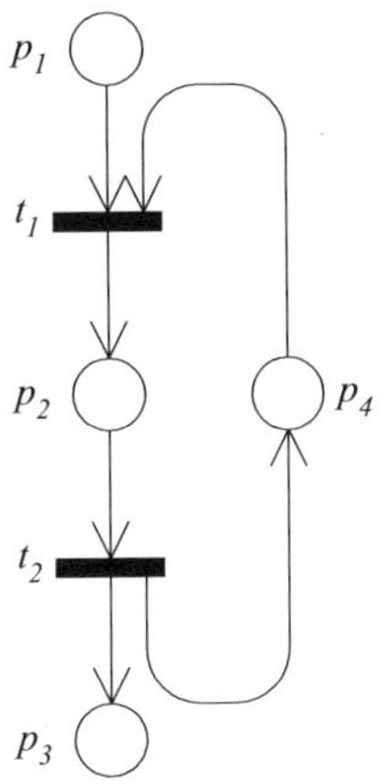

4 places:
p_1 = aircraft available for service
p_2 = aircraft maintenance operation being performed
p_3 = aircraft maintenance complete and available for use
p_4 = hanger available

2 transitions:
t_1 = aircraft moved into hanger
t_2 = aircraft moved out of hanger

Figure 7.1.2-1 Petri net for aircraft maintenance at a hangar

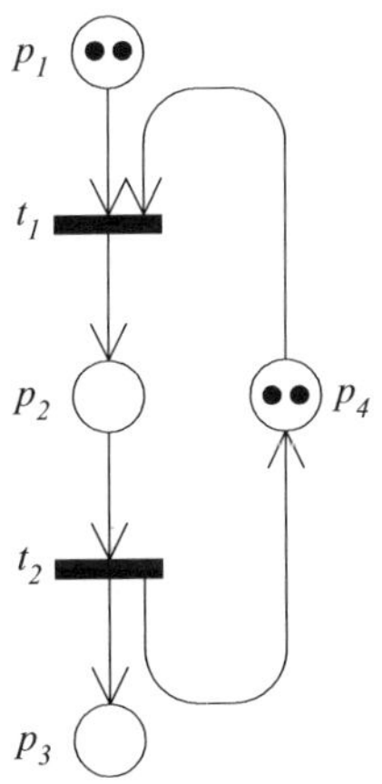

4 places:
p_1 = aircraft available for service
p_2 = aircraft maintenance operation being performed
p_3 = aircraft maintenance complete and available for use
p_4 = hanger space available

2 transitions:
t_1 = aircraft moved into one of the spaces in hanger
t_2 = aircraft moved out of hanger

Figure 7.1.2-2 Petri net for aircraft maintenance at a large hangar for two aircrafts

7.1.3 System dynamics

System dynamics modelling methods are commonly used in determining macro variations of subsystems due to changing operation parameters. The high-level system is modelled quantitatively and decomposed into smaller subsystems that can be more readily described in terms of mathematical behaviour (Sterman, 2000).

No matter how complex a system is, its system dynamics model can be represented by a causal diagram of appropriate level. A causal diagram consists of variables connected by arrows denoting the causal influences among the variables. Causal links are

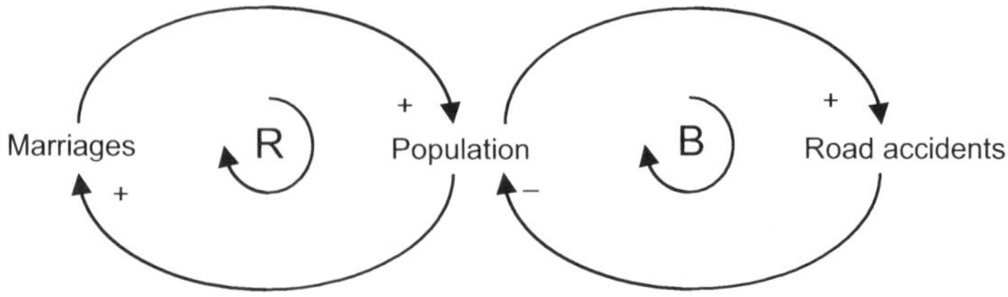

Figure 7.1.3-1 Causal loop diagram of chicken population, used to demonstrate systems dynamics

assigned a polarity, either positive (+) or negative (–), to indicate how the dependent variable changes when the independent variable changes. The characteristics of the links are:

- *A positive link*. This means that if the cause increases, the effect increases, and if the cause decreases, the effect decreases as well.
- *A negative link*. This means that if the cause increases, the effect decreases, and if the cause decreases, the effect increases as well.

The most complex behaviours in any system usually arise from the interactions (feedbacks) among the components of the system. Feedbacks are returning links that form positive and negative feedback loops. Positive feedbacks are self-reinforcing and amplify whatever is happening in the system. To illustrate this behaviour, Figure 7.1.3-1 represents a causal loop diagram in which more people in a city (population) result in more marriages, which in turn give new births and add to the population. This loop, which is self-reinforcing, is given the polarity identifier 'R'.

The second loop represents the negative feedback. It is self-correcting in that it counteracts changes indicated in the first loop. In this example, as the population grows, there is more road traffic and hence more road accidents in the city, which lead to more unnatural deaths that will, in turn, bring the population down. The 'B' in the centre of the loop denotes this balancing feedback.

Note that the loop identifier circulates in the same direction as the loop to which it corresponds.

System dynamics models incorporated within computer simulations can be used to investigate systems that are characterised by non-linearity, delay and feedback processes. This simulation outcome can increase understanding of quantitative and qualitative predictions of systems behaviour. Figure 7.1.3-2 presents a more complicated example of a system dynamics model of a hybrid manufacturing resources planning (MRP) and Kanban supply chain system. The model presents the stock and flow and decision structure at a high level (not at the level of the individual equations). The company is defined by the large round-edged rectangle containing smaller round edged rectangles which denote organisational subunits, policies or decision rules.

To understand the model, we start from the right-hand side. First, the 'Customer Order Rate' is the stochastic demand received by the manufacturing company; it is outside of the system's boundary (exogenous variable). The company aims to hold a

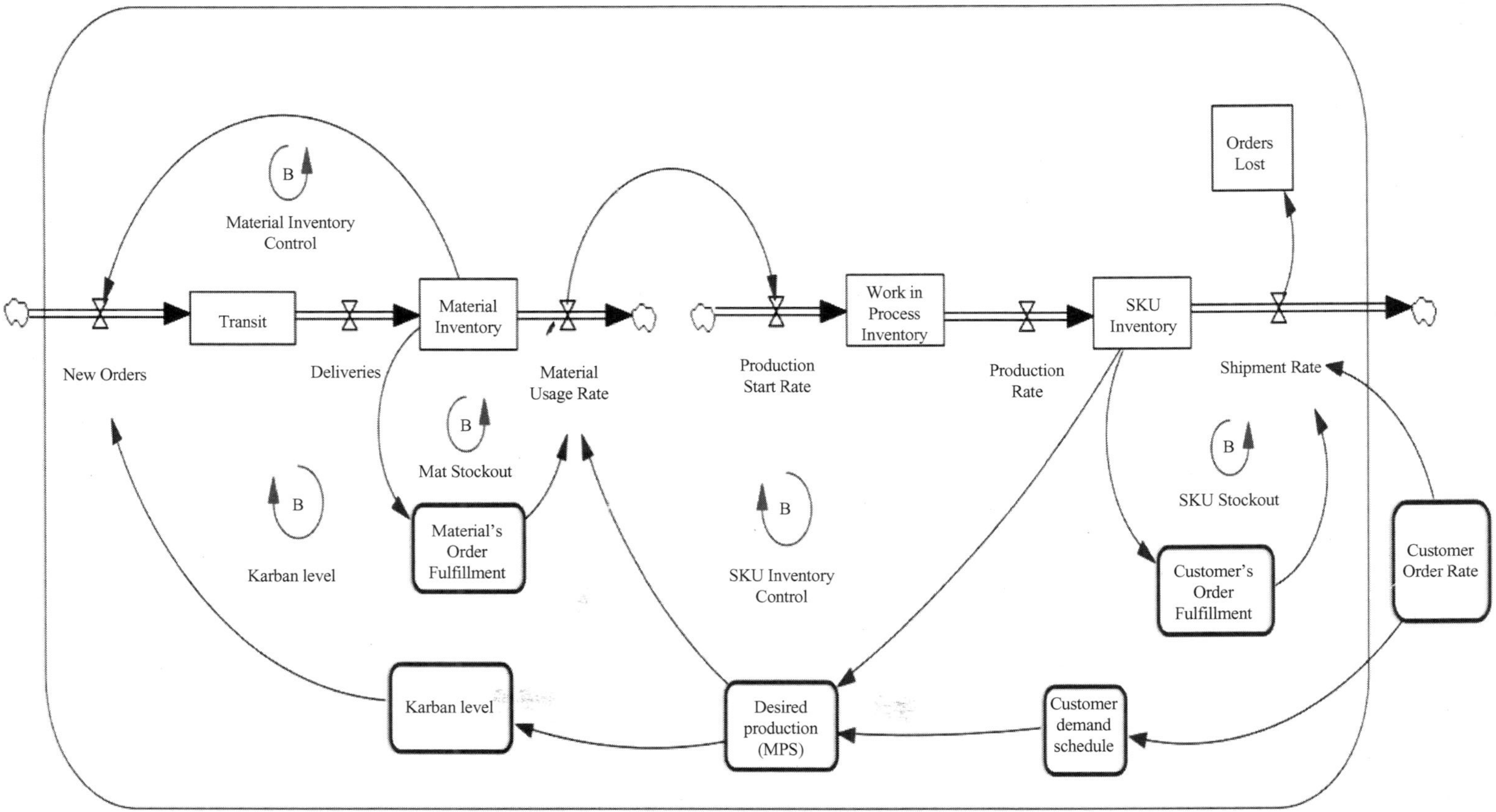

Figure 7.1.3-2 The system dynamics model of the hybrid (MRP-Kanban) system

'SKU Inventory' (i.e. its stock of finished goods) so orders can be filled as they arrive. It is assumed that customers are delivery-sensitive (i.e. that they will seek other sources of supply when orders cannot be filled upon request; thus these customers are lost).

Production takes time. The stock of work-in-process (WIP) is increased by 'Production Start Rate' and decreased by 'Production Rate'. 'Desired Production', which represents the master production schedule (MPS), drives the rate of production starts based on the demand forecast and the inventory position of final product. However, the Production Start Rate is also affected by the availability of raw materials; it is assumed that production can only start if there is sufficient stock of materials. The company is required to order a sufficient amount of materials to keep its 'Materials Inventory' at the appropriate level. No other restriction (e.g. machine breakdown or capacity constraints, either from labour or capital) is assumed to affect production.

Material inventory is increased by 'Deliveries' and decreased by 'Material Usage', which depends on the production schedule (determined by MPS). Materials are replenished under a Kanban electronic system. However, the MPS also drives the 'Kanban level', giving place to the hybrid MRP/Kanban system.

The 'Transit' of parts represents the stock of 'Pieces to be delivered' by the raw materials supplier. Transit takes time (i.e. lead time). This stock is increased by 'New orders' and decreased by 'Deliveries' when they are physically delivered to a facility's warehouse. New orders are generated based on the Kanban level and the inventory position of materials. Note that new orders are triggered depending on the materials inventory and number of Kanbans the firm desires to hold.

This model includes five important negative feedbacks:

- The *SKU Stockout* loop regulates shipments as inventory varies. If inventory is inadequate, some items will be out of stock and orders will be lost. In the extreme, shipments fall to zero when there is no inventory.
- The *SKU Inventory Control*' loop adjusts production start rate to move the levels of SKU inventory towards their desired levels.
- The *Material Stockout* loop regulates material usage rate as inventory varies. If inventory is inadequate, some items will be out of stock and Production Start Rate will be affected. In the extreme, material usage falls to zero when there is no inventory. That means an out of materials and production of desired product won't be started.
- The *Kanban level* loop regulates the number of Kanbans to be held to trigger new orders. It is based on MPS.
- The *Material Inventory Control* loop adjusts new orders to maintain optimal materials inventory.

The high-level model is expanded to lower levels, where more detailed feedback loops are created to represent the system. For example, in the detail level on raw materials usage, some raw materials are shared among several products. After inserting some typical parameter values, the model is numerically integrated; responses are shown in Figure 7.1.3-3.

According to Figure 7.1.3-3, the SKU inventory level is quite high. However, it should not be reduced because doing so may affect the raw materials inventory, which shows a high level of fluctuation.

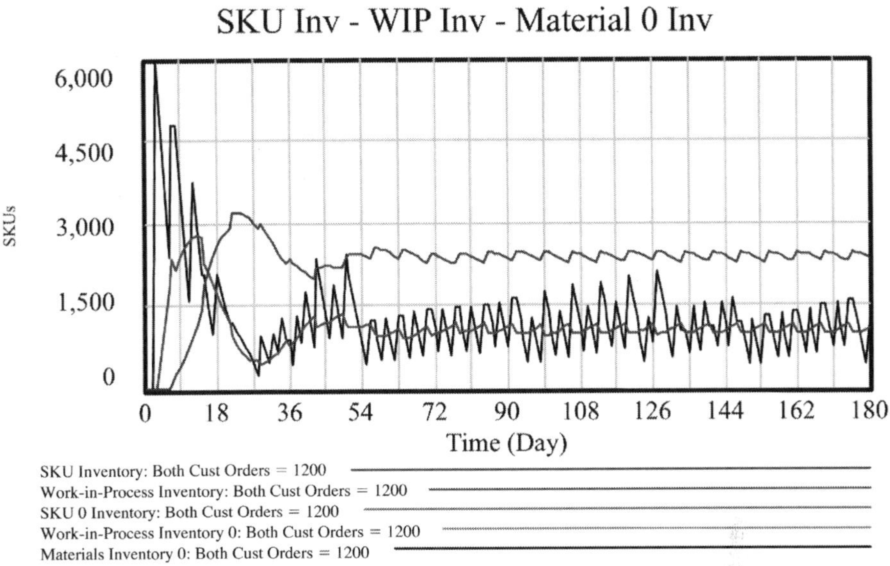

Figure 7.1.3-3 Materials, WIP and SKUs inventories at constant demand rate

A system dynamics model provides quantitative analysis of the behaviour of systems with stock and flow characteristics. It takes a lot of effort to create and is time-consuming to debug. However, once it is validated, the possibility of scenario investigation is endless.

7.1.4 Monte Carlo simulation

Support-system design decisions based on the probability distributions of present worth associated with various system alternatives are easily understandable and presentable to stakeholders. One obvious question is: How are such distributions developed? One simple answer is to consult a few experts. From data obtained, the probabilities can be estimated. This approach is not generally used as it is difficult to find persons knowledgeable enough about all aspects of a support system to make reliable estimates of the probability distributions.

The more common approach is to identify the significant parameters that will affect the operations of the support system. People in key areas of experience in the fields represented by the various parameters are asked to estimate the probability for a parameter about which they have knowledge. Plant operations will consider operating costs, for example marketing group the market share and finance group the cost of capital.

Once probability distributions are prepared for each of the significant parameters (i.e. cost, income, life, inflation, interest rate etc.) the overall profit of the support contract is defined by establishing a linkage between each system parameter. Simulation using Monte Carlo techniques is then applied to produce the probability

Table 7.1.4-1 O & M costs of alternatives A and B and their probabilities

Alternative	O & M cost ($)	Probability
A	2,000	1/6
	3,000	1/2
	5,000	1/3
B	12,000	1/6
	25,000	1/3
	30,000	1/3
	40,000	1/6

distribution of the investment's present worth, annual equivalent or other basis for comparison.

Monte Carlo is the name given to a class of simulation approaches to decision making in which probability distributions describe certain economic elements. In many of these uses, an analytical solution is not possible because of the way the probabilities must be evaluated. An example follows to illustrate the Monte Carlo process.

A manufacturer wishes to enter a bid to supply a special instrument to a client. Two alternatives are being considered. Alternative A has a high acquisition cost of $70,000 and low operation and maintenance (O & M) costs because of its high degree of automation. Alternative B has a low acquisition cost of $20,000 and high O & M costs. Although the acquisition cost of each alternative is known, the annual O & M costs are uncertain. The following data have been collected for analysis:

The rate of return is 10% per annum.

The O&M costs of alternatives A and B are given in Table 7.1.4-1

To add to the complexity, the contract duration is uncertain and is estimated to be either 1, 2 or 3 years (Table 7.1.4-2).

Table 7.1.4-2 Possibilities of contract duration

Contract duration (years), *n*	Probability that the duration is *n* (years)
1	0.25 (i.e., 1/4)
2	0.50 (i.e., 1/2)
3	0.25 (i.e., 1/4)

Table 7.1.4-3 Contract duration generator assignment

Simulated value (years)	Probability of outcomes	Assignment of outcome
1	1/4	HH
2	1/2	HT or TH
3	1/4	TT

Table 7.1.4-4 O & M costs of alternatives A and B and their probabilities

Alternative	O & M cost ($)	Probability	Assignment of dice outcomes
A	2,000	1/6	1
	3,000	1/2	2, 3, or 4
	5,000 .	1/3	5 or 6
B	12,000	1/6	1
	25,000	1/3	2 or 3
	30,000	1/3	4 or 5
	40,000	1/6	6

Table 7.1.4-5 Simulation table for special instrument bid

(a)	(b)	(c)	(d)	(e)	(f)	(g)	(h)	(i)	(j)	(k)
HH	1	77,000	22,000	2	3,000	80,000	3	25,000	47,000	33,000
HT	2	40,334	11,524	4	3,000	43,334	3	25,000	36,524	6,810
TH	2	40,334	11,524	1	2,000	42,334	6	40,000	51,524	−9,190
TH	2	40,334	11,524	5	5,000	45,334	2	25,000	36,524	8,810
TT	3	28,147	8,042	2	3,000	31,147	1	12,000	20,042	11,105
HH	1	77,000	22,000	3	3,000	80,000	5	25,000	47,000	33,000
TH	2	40,334	11,524	6	5,000	45,334	4	30,000	41,524	3,810
HT	2	40,334	11,524	1	2,000	42,334	3	25,000	36,524	5,810
TT	3	28,147	8,042	3	3,000	31,147	6	40,000	48,042	−16,895
TH	2	40,334	11,524	6	5,000	45,334	4	30,000	41,524	3,810
					Mean	48,630			41,623	7,007

(a) Outcome of coin toss; (b) contract duration; (c) annual equivalent value of alternative A; (d) annual equivalent value of f alternative B; (e) outcome of dice for A; (f) O&M cost of A; (g) total annual equivalent cost of A; (h) outcome of dice for B; (i) O & M cost of B; (j) total annual equivalent cost of B; (k) difference in cost: A–B.

We first illustrate the simulation by a mechanical means to generate the random numbers. If the probability is a multiple of 1/4, two coins can be used. If the probability is a multiple of 1/6, one die used. Hence, for contract duration, the 4 different outcomes can be generated by tossing 2 coins as shown in Table 7.1.4-3.

The O&M costs can be assigned with dice outcomes as shown in Table 7.1.4-4.

We then set up the simulation table as shown in Table 7.1.4-5.

The coins and dice are tossed for every time when a selection due to probability is required to be generated. For example, in row 1 of Table 7.1.4-5, for contract duration, the coins are tossed and two heads are obtained; according to Table 7.1.4-4, duration of 1 year is assigned. Going across Table 7.1.4-5, at column (e), the outcome of a dice throw is 2. According to Table 7.1.4-2, a value of $3,000 is assigned to alternative A O&M cost. The remaining columns of Table 7.1.4-4 are made similarly.

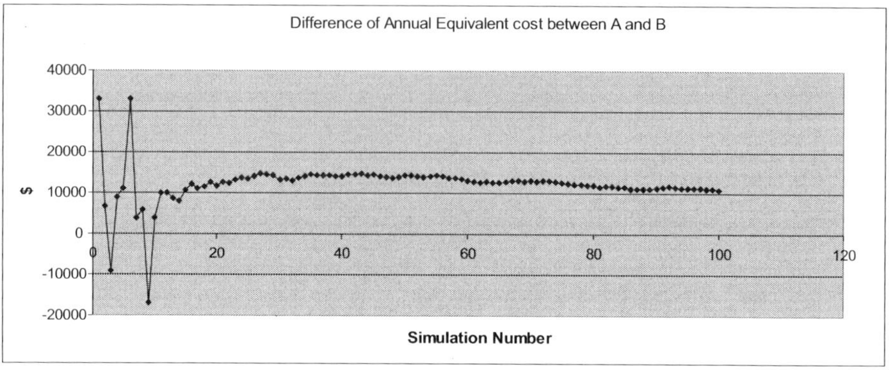

Figure 7.1.4-1 Plots showing convergence of annual equivalent cost with increasing runs

After 10 trials, the average annual equivalent cost for alternative A is $48,630 and the average annual equivalent cost for alternative B is $41,623. The annual equivalent cost difference is $7,007. This means that B is cheaper overall. Monte Carlo simulation requires a sufficiently large sample to be taken. After 100 simulation runs, a possible outcome can be plotted as in Figure 7.1.4-1.

The difference of the annual equivalent cost between the two alternatives can be seen to become steady after around 80 runs to be $9,000 approximately. This means alternative B has a better chance to produce a lower cost solution than alternative A.

The model can be simulated on Excel similar to Table 7.1.4-4. The device used to create the random numbers on a digital computer is the random number generator function 'rand()'. This generator is usually a program that returns values from a uniform distribution of the intervals 0.0 to 1.0. Instead of column (a), the random function is inserted into the cells. Column (b) is then replaced by the function:

if ({col (a)} < 0.25, 1; if ({col (a)} < 0.75, 1, 2))

For columns (c) and (h), two independent "rand()" columns are required. Columns (c) and (h) are replaced by functions:

- if ({col (c) rand()} < 1/6, 2000; if ({col (c) rand()} < 2/3, 3000, 5000))
- if ({col (h) rand()} < 1/6, 12000; if ({col (h) rand()} < 1/2, 25000,, if ({col (h) rand()} < 5/6, 30000, 40000)))

When the rows are copied, the simulation effect can be generated. The average annual equivalent cost can easily be calculated from the last column.

7.1.5 Discrete system simulation

Discrete system simulation is becoming more popular due to its easier modelling construct. IT is useful for statistical concepts, particularly those related to sampling distributions and random occurrences. A discrete system is one that changes state abruptly at discrete points in time. Most systems encountered in operations research and management science (e.g. studies of communication or transportation) are discrete.

A simulation model contains a physical model and a logic model. The physical model is a physical replica of the real system, which is built basically for vision requirement and makes the model clear to tell and easily to be read. All the models and functions are logical model, but the sequence is the key part of the system because it is required to simulate the real situation.

From a practical viewpoint, simulation is the process of designing and creating a computerised model of a real or proposed system for the purpose of condition numerical experiments that will give us a better understanding of the behaviour of that system for a given set of conditions.

We use a simple discrete simulation model of a train line in Melbourne. Figure 7.1.5-1 shows the complete metropolitan route map. The two lines that are being modelled for this example Hurstbridge and Epping lines, are highlighted.

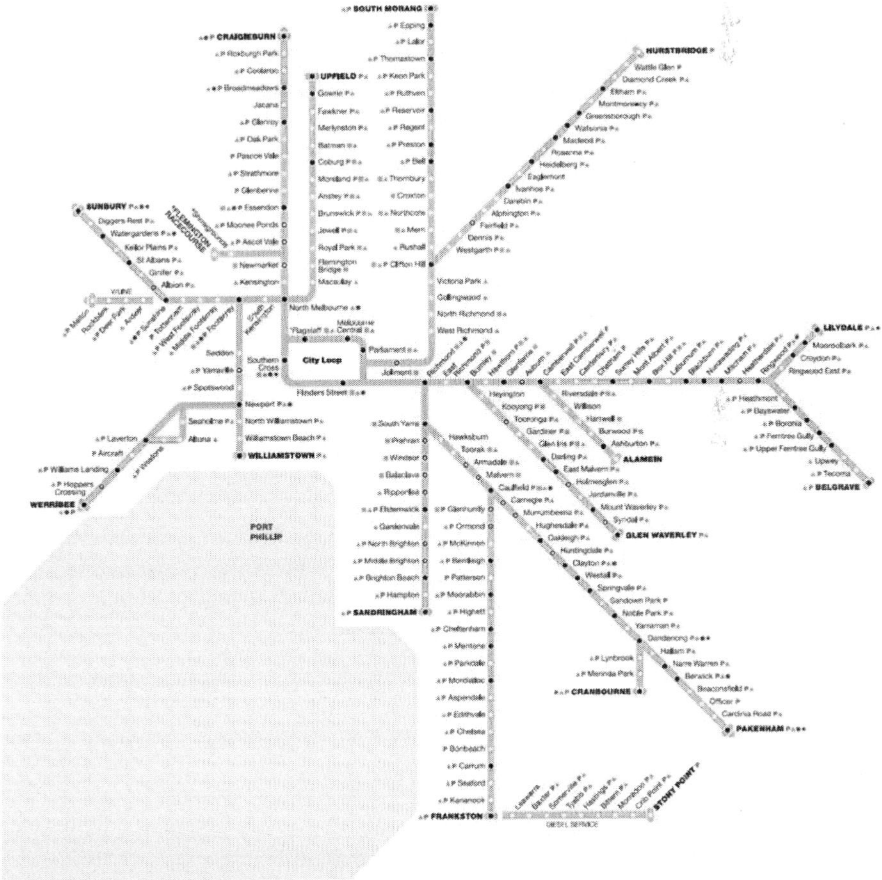

Figure 7.1.5-1 The Melbourne Metropolitan train network

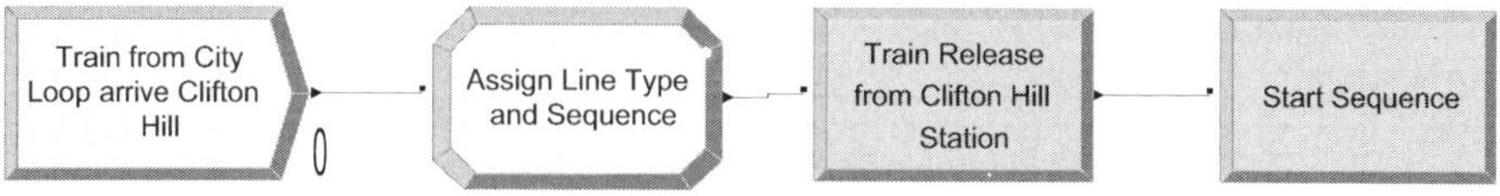

Figure 7.1.5-2 Train arrival Clifton Hill

Figure 7.1.5-2 shows that the train from City Loop arrives at Clifton Hill Station, where it will go onto one of the two branch routes. The probability of going into either branch can vary. In the simulation model, after assigning the line type, the train entity in the model will follow one of the defined simulation sequences.

'Sequence', which refers to an ordered list of objects, is a definition of control within the simulation environment. A sequence contains elements; the number of those elements is called the 'length of the sequence'. The movement of entity in this model is according to the sequence, which is shown in Table 7.1.5-1.

The sequence of Line 1 is called Line 1 Process Plan; it is indicated in Figure 7.1.5-3 by the green arrowhead. It is assumed that this sequence begins when each Line 1 is released from Clifton Hill Station. According to Line 1 Process Plan, Line 1 goes into Preston Station as its Stop 1, goes through Preston Process, and is released at Route from Preston. After that, it goes into Epping Station as its Stop 2, goes through Epping Process, and is released again at Route from Epping. Finally it goes into the Exit Line Network Station as its Stop 3 and gets disposed from train lines (i.e. it leaves the train system).

The sequence of Line 2 is called Line 2 Process Plan; it is shown by the blue arrowhead in Figure 3-10. Its sequence begins when each Line 2 is released from Clifton Hill Station, just as with Line 1. The Line 2 goes to Ivanhoe Station as its Stop 1, then goes through Ivanhoe Process, and is released at Route from Ivanhoe. The Eltham Station is its Stop 2; it will be released at Route from Eltham; and then it will go to Exit Line Network Station as its Stop 3. After that, it is disposed (leaves the system).

When a train (i.e. an entity) must stop (i.e. delay its entry process) and wait for a train that is in process (i.e. a primed entity), the waiting train is said to be 'in queue'. When an entity cannot move on (perhaps because it needs to seize a unit of a resource that is tied up by another entity) it needs a place to wait, which is the purpose of a queue. It is necessary to decide as part of simulation modelling how to handle an entity arriving at a queue that is already full.

After setting up the parameters of the model such as train schedule, the model can be simulated to access the behaviours of interest. For example, the number of

Table 7.1.5-1 Train routings

Line No.	Station	Station	Station
1	Clifton Hill	Preston	Epping
2	Clifton Hill	Ivanhoe	Eltham

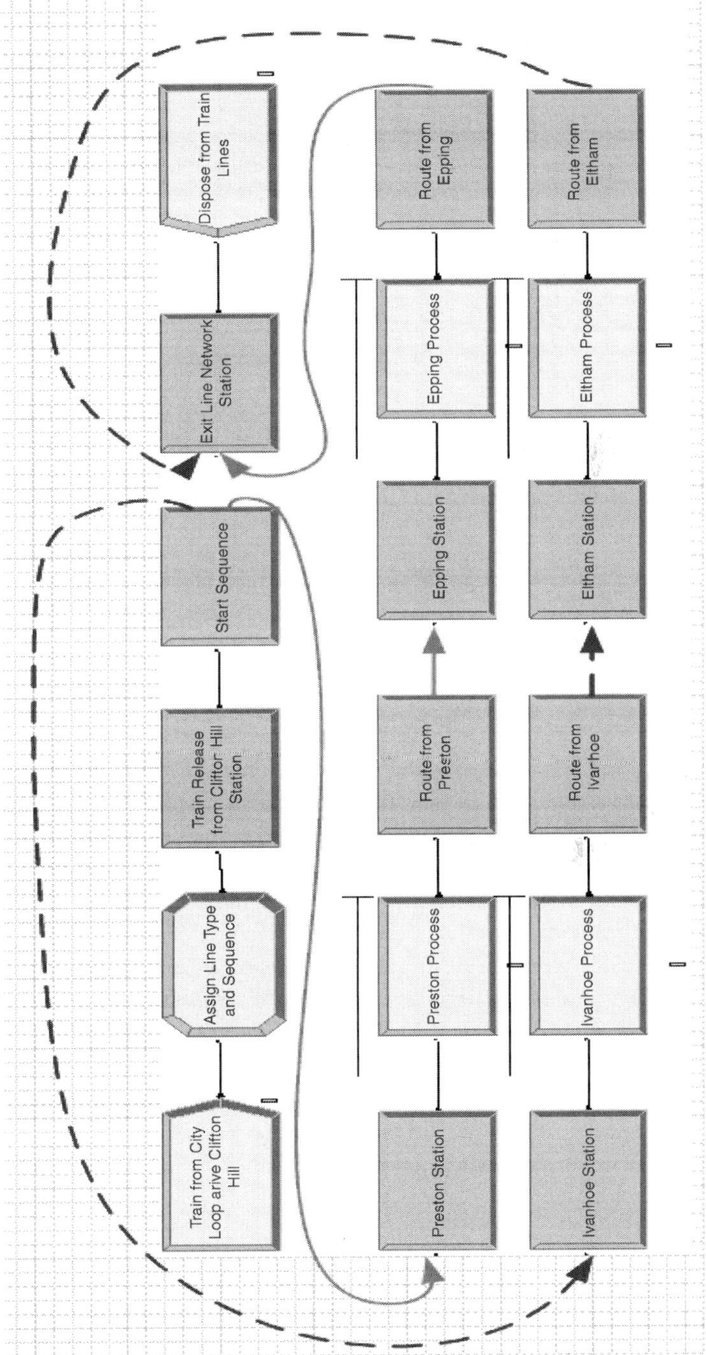

Figure 7.1.5-3 Sequence model

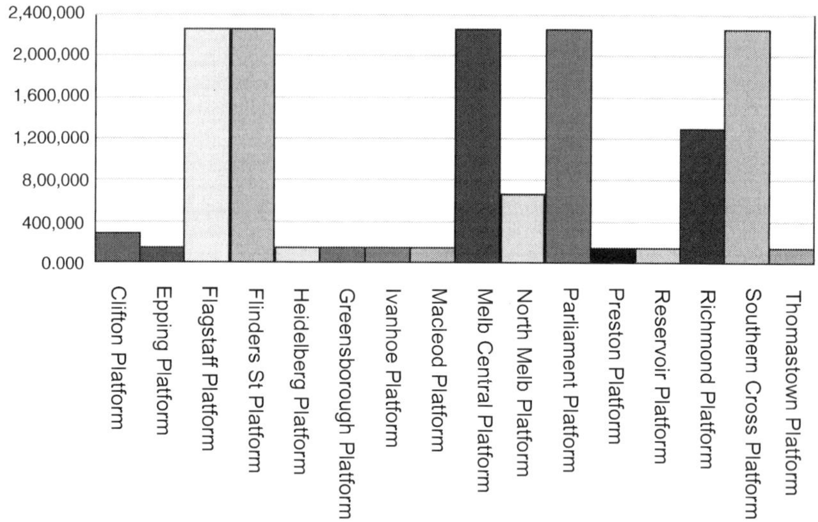

Figure 7.1.5-4 Simulated number of passengers passing through stations

passengers through different stations on the lines can be modelled easily, as shown in Figure 7.1.5-4.

7.2 Vehicle routing

Physical distribution activities in supply chains are of major concerns to business managers, operators and owners. Customer satisfaction in distribution supply chains depends on fast, reliable means of moving the goods, particularly for time-critical distributions such as dairy-supply chains. An example of a time-critical distribution supply chain is the operation requirements of a dairy distribution company.

Because most dairy products are perishable commodities, orders are scheduled by a dairy company on a just-in-time schedule. Therefore, a company's orders are packed overnight at a cold-storage facility with the help of a third-party logistics provider.

To reduce the distribution lead-time (defined as the period of time that elapses between placing an order and receipt of goods), the efficiency and effectiveness of fleet management are of paramount importance. Vehicle routing planning (VRP) is one of the critical factors in fleet management. VRP has wide applicability and economic importance in determining efficient distribution strategies to reduce operational costs in distribution systems.

Figure 7.2-1 demonstrates the basic process of VRP. Vehicles, customers, depots and routes are considered, but various common restrictions for every factor must also be considered. VRP can be classified to a certain extent as relating either to the vehicles or to the customers. Not all of the restrictions may apply in any particular case. Therefore, conditions could be listed in every factor. Some factors in this case are:

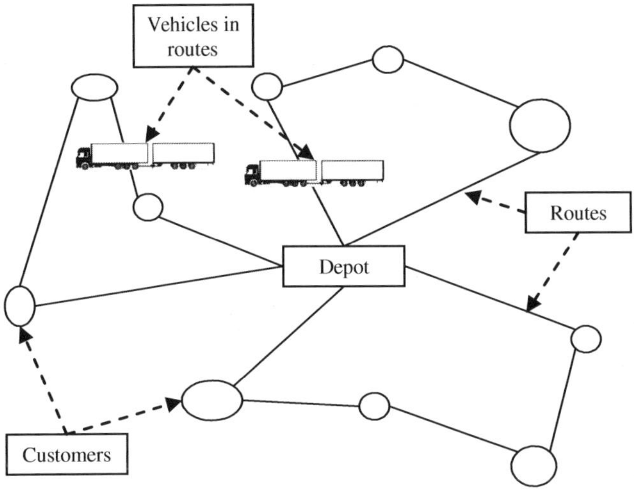

Figure 7.2-1 The basic routing problem

- Depots (number, location).
- Vehicles (capacity, costs, time to leave, driver rest period, type and number of vehicles, max time).
- Customers (demands, hard or soft time windows, pickup and delivery, accessibility restriction, split demand, priority).
- Route Information (maximum route time or distance, cost on the links).

Several objectives could be adopted in order to meet the above conditions and requirements, such as:

- Minimise the total travel distance (or time).
- Minimise the number of vehicles used.
- Minimise the combination of number of vehicles and total distance (or time).

As the objective of VRP is to minimise the cost of distribution, the first objective corresponds to the minimisation of total travel distance; the second objective corresponds to the minimisation of the numbers of vehicles used in distribution; and the third objective is to minimise the combination of objectives 1 and 2.

Many different variants of this problem have been formulated to provide suitable applications to a variety of real-world cases. VRP solutions not only consider constraints of a problem but also take into account the different aspects of its nature, such as its dynamism, time dependency and/or stochastic aspects.

There are two main approaches for VRP. Both are exact and heuristic. Exact approaches include branch and bound, cutting planes, network flow, dynamic planning etc. A heuristic approach uses an heuristic algorithm to determine an optimal or near-optimal route. Typical heuristic algorithms include constructive, sweep, two-phase, tabu search, genetic, ant colony optimisation etc. To illustrate how an heuristic algorithm works, we use a relatively simple algorithm in a worked example.

7.2.1 Vehicle routing problem defined

To solve the VRP, it is necessary to define clearly the data and constraints in the logistic system. Some major criteria such as number of trucks, location of hub and customers and delivery distance are essential starting information (see Table 7.2.1-1).

Table 7.2.1-1 Definition of information in a VRP formulation

Item	Description
Product and delivery unit	The demand is dairy product and the delivery unit is pallet. The amount of demand for customers is random, from 1 to 10 pallets.
Trucks	Factors about trucks relate to truck size, truckload, number of trucks used and truck-using time.
Truck size	Suppose there are three truck sizes in the distribution fleet: small, medium and large. Different sizes of trucks have their own truckloads, which is beneficial to deliver products to different customers due to the product demand and delivery distance.
Truckload	The truckload is random with logical sense. A small truck can deliver a maximum 20 pallets dairy products a day; medium truck can afford 30 pallets; and the large-truck truckload is 50 pallets. Sometimes a truck may not need full upload, based on the delivery task and routing; but sometimes more products can be uploaded, in excess of the amount of current task, to ensure the availability of temperate new added task.
Number of trucks	Suppose that there is only one truck for every truck size. During the vehicle routing and distribution-arrangement planning, different sizes of trucks may use different routes due to varying distribution amounts, economic factors etc.
Utilisation time	Every size does have a use-time limit. Hence when any truck finishes its tasks it can directly go back to the hub and then either stop its daily trips, refill to continue the delivery schedule or continue delivering (or other temporary tasks) according to the factual vehicle routing arrangement. In addition, some size trucks may be used less, according to distribution tasks and economic factors. For example, when limited customers requiring small amounts of product are relatively far from the hub, a small truck would be a better choice than a large truck.
Location	Information about locations in this research is all taken from the Internet. The distances of hubs and stores are all taken from Google Maps.
Location of hub	In this research, suppose National Foods has only one distribution hub, located in 1 Gower St, Preston, VIC.
Location of customers/stores	There are 20 Woolworth's stores near the distribution hub to be used as samples of customers/store locations. See Table A1 for details of location and store information.
Distance	The addresses of hubs and customers are all taken from Google Maps, as are the distances between hubs and stores and distances between stores (see Table A2). The unit of distance is kilometres.
Supposed conditions	In practice, some factual situations may occur during product distribution. Hence some supposed conditions are set to before planning vehicle routing.

Table 7.2.1-4 **Distance savings**

	1	2	3	4	5	6	7	8	9	10	11	12	13	14	15	16	17	18	19	20	
1		4.3	3.5	2.8	5.4	6.8	4.6	2.2	2.8	2.5	1.3	4.6	6	2.2	7.4	5.5	3.2	1.7	2.3	2.5	**1**
2	4.3		4.2	−0.2	−0.3	2.8	5.1	1	4.8	1.1	3.3	5.1	2.8	4	2.7	5.3	5.2	0.5	4.1	2.6	**2**
3	3.5	4.2		0.8	−0.5	2.5	5.6	3.4	5.9	9.2	5.4	1.6	2.5	1.1	2.4	6.5	6.2	0.3	6.7	−0.6	**3**
4	2.8	−0.2	0.8		7.9	2.1	1.2	4.2	1.8	12.9	0.9	2.3	3.6	−1.8	9.6	1.9	0.7	8.7	2.4	0.1	**4**
5	5.4	−0.3	−0.5	7.9		2.6	−0.8	16.5	8.8	17.1	7.8	2.1	3.4	0.8	7.8	14.6	4.1	21.4	21.3	−0.1	**5**
6	6.8	2.8	2.5	2.1	2.6		2.3	0.4	−15.3	9.9	−0.5	14.3	15.7	−15.8	15.2	−13.3	−8.6	1.2	−21	12.2	**6**
7	4.6	5.1	5.6	1.2	−0.8	2.3		3	10.6	12.5	6.6	−2.2	0.1	5.4	−0.4	11.3	12.6	−0.1	0.2	−3.4	**7**
8	2.2	1	3.4	4.2	16.5	0.4	3		11.7	20	10.7	−0.3	2	8	7.9	14.3	9.3	17.3	18.3	−1.5	**8**
9	2.8	4.8	5.9	1.8	8.8	−15.3	10.6	11.7		20.5	16	−14.6	−12.3	18.3	17.4	21.7	18.8	11.7	17.3	−15.8	**9**
10	2.5	1.1	9.2	12.9	17.1	9.9	12.5	20	20.5		18.9	8.8	11.1	16.3	−12.2	19.6	18.8	19.2	19.9	7.6	**10**
11	1.3	3.3	5.4	0.9	7.8	−0.5	6.6	10.7	16	18.9		−2.1	0.2	12.5	0.8	18.1	13.7	12.2	16.8	−3.3	**11**
12	4.6	5.1	1.6	2.3	2.1	14.3	−2.2	−0.3	−14.6	8.8	−2.1		20	−12.2	16.4	−9.8	−5	−1.3	−17.5	16.4	**12**
13	6	2.8	2.5	3.6	3.4	15.7	0.1	2	−12.3	11.1	0.2	20		−13.1	17.8	−10.7	−5.9	4.2	−18.3	18.7	**13**
14	2.2	4	1.1	−1.8	0.8	−15.8	5.4	8	18.3	16.3	12.5	−12.2	−13.1		2.2	19.6	19.7	9.5	10	−14.9	**14**
15	7.4	2.7	2.4	9.6	7.8	15.2	−0.4	7.9	17.4	−12.2	0.8	16.4	17.8	2.2		2.3	−6.5	8.6	2.3	18	**15**
16	5.5	5.3	6.5	1.9	14.6	−13.3	11.3	14.3	21.7	19.6	18.1	−9.8	−10.7	19.6	2.3		19.5	14.1	21.3	−15	**16**
17	3.2	5.2	6.2	0.7	4.1	−8.6	12.6	9.3	18.8	18.8	13.7	−5	−5.9	19.7	−6.5	19.5		2	12.7	−4.2	**17**
18	1.7	0.5	0.3	8.7	21.4	1.2	−0.1	17.3	11.7	19.2	12.2	−1.3	4.2	9.5	8.6	14.1	2		18	0.7	**18**
19	2.3	4.1	6.7	2.4	21.3	−21	0.2	18.3	17.3	19.9	16.8	−17.5	−18.3	10	2.3	21.3	12.7	18		18.2	**19**
20	2.5	2.6	−0.6	0.1	−0.1	12.2	−3.4	−1.5	−15.8	7.6	−3.3	16.4	18.7	−14.9	18	−15	−4.2	0.7	18.2		**20**
	1	**2**	**3**	**4**	**5**	**6**	**7**	**8**	**9**	**10**	**11**	**12**	**13**	**14**	**15**	**16**	**17**	**18**	**19**	**20**	

7.2.2 Assumptions

The solution is illustrated using an example. In Table 7.2.2-1, 20 store locations are identified. These stores are served by a central hub. It is assumed that all sizes of trucks originally locate in the hub and then leave the hub with planned routes.

Table 7.2.2-1 Store information

No.	Location	Address	Distance from hub (km)	Suggested route	Quantity of goods in trip
1	Reservoir	820/822 Plenty Rd	3.7	Albert St and Plenty Rd	7
2	Preston	10 Clinch Ave	2.9	Murray Rd	1
3	Preston	Cnr Plenty Road & Dundas Street	3.4	Bell St	6
4	Heidelberg	451 Lower Heidelberg Road	4.5	Bell St and Banksia St	6
5	Camberwell	57 Station Street	12.6	Burke Rd	7
6	Bundoora	69-85 Plenty Road	7.6	Albert St and Plenty Rd	4
7	Coburg	Cnr Munro & Louisa Streets	6.6	Bell St	10
8	Kew	7 Walpole Street	9	Grange Rd	5
9	Carlton	368/380 Lygon Street	11.2	Nicholson St	2
10	Eltham	7 Arthur Street,	14.8	Lower Plenty Rd and Main Rd	7
11	Fitzroy	243 Smith Street	8.5	Station St and Heidelberg Rd	10
12	Mill Park	Childs Road	11.4	Plenty Rd	2
13	Mill Park	Cnr Childs Road & Redleap Avenue	11.9	Plenty Rd	4
14	Abbotsford	313 Victoria Street	10.1	Station St	10
15	St Helena	Cnr St Helena & Aquaduct Roads	16.3	Plenty Rd	7
16	Melbourne	Cnr Lonsdale & Swanston Streets	12.7	Nicholson St	7
17	Maribyrnong	Rosamond Rd	17.7	Maribyrnong Rd	8
18	Balwyn	Whitehorse Road	12.3	Burke Rd	1
19	Toorak	459-465 Toorak Road	15	Station St	6
20	South Morang	Plenty Road	12.5	Plenty Rd	7

Figure 7.2.1-1 Map of hub and Woolworth's store locations

This assumption is essential for every distributor during the daily vehicle-routing planning. Generally all the trucks originally park in the hub to await the new tasks.

This assumption affects truck-utilisation time. Suppose that truck S and truck L run just once for the daily distribution and truck M runs twice.

This assumption is based on the real situation in which the trucks are out of the hub with enough product. They can keep running and distributing the products to other customers if they get the appropriate routing sequence from the distribution management centre immediately. This assumption can support and test the vehicle-routing planning from the distributor if the vehicle routing can be rearranged quickly in light distribution feedback and the degree of the planning feasibility.

Table 7.2.1-1 shows the stores and their distances from the hub, computed using Google Maps. Markers A–J are the locations of stores 1–9. Due to traffic constraints (e.g. one-way streets, road configurations etc.) the distances are longer than the direct line. The quantity of goods in the trip being planned is also listed. Figure 7.2.1-1 shows the locations of some of the stores. In addition, the green mark in the map is the place where the hub locates.

The hub will be served by three trucks: S, M and L. The capacities of the trucks are shown in Table 7.2.2-2.

Table 7.2.2-2 Truck capacities

Truck	Capacity
S	20
M	30
L	50

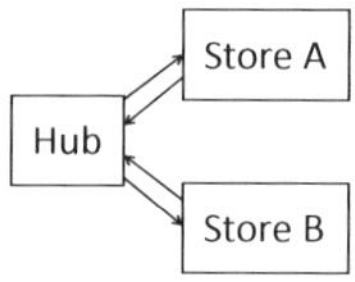

Figure 7.2.2-1 Trips without considering savings

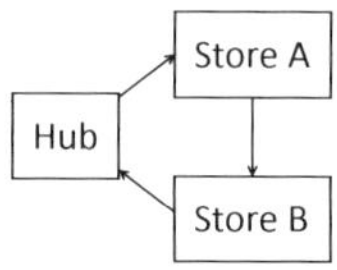

Figure 7.2.2-2 Trips after considering savings

In practice, the number of store locations and quantity of goods demands more than one truck or route. Therefore, a procedure to progressively determine the best possible set of routes is required.

7.2.3 Savings method

A savings algorithm is generally based on the saved distance to every pair of customers. In Figure 7.2.2-1, the original route is from hub to Store A; it then goes back to the hub, where the truck will load more goods for a return route to Store B. The savings concept is to travel in a way that removes the return trip to the hub, as shown in Figure 7.2.2-2 (Graham, 2009).

The savings distance is given by:

$$D_{saved} = D_{A\text{-}hub} + D_{hub\text{-}B} - D_{A\text{-}B}$$

In addition to knowing the distance between hub and stores, it is also necessary to measure the distances between stores and stores. Based on the information in Table 7.2.2-2, and additional investigation with Google Maps, a travelling distance table is tabulated in Table 7.2.3-1. It is noted that location 0 is the hub. The table is

Table 7.2.3-1 Store information

Stores	Distance	Quantity of goods	Trucks
1	3.7	7	L
2	2.9	1	S
3	3.4	6	M

triangular (i.e. the lower half of the table below the diagonal empty elements is a reflection of the upper half). It is normal practice to leave the lower-half cells empty; however, for the sake of completeness, we have included the lower-half values. Table 7.2.3-1 forms the basis for the routing assignment. We use the assumption outlined previously (i.e. all three trucks are located at the hub originally and leave the hub with a planned route to the stores, loaded with all dairy orders for that route). The algorithm has the following steps:

1. From Table 7.2.3-1, compute the distance-savings table (Table 7.2.3-2).
2. Pick up three closest distances from hub to store as a possible next stop after the start-point hub; stores 1, 2 and 3 are matched. Due to the demand requirement, store 2 requires least quantity; hence, the smallest truck S is allocated to this task. Store 1 has the biggest order; hence, the largest truck L is allocated to this task. Truck M gets the task for store 3, which needs moderate quantity delivery (Table 7.2.3-2).
3. From store 1, the next stop for truck L is determined by the largest savings for store 1. From Table 7.2.2-1, store 15 gives the largest savings; hence, truck L is allocated to this task. Truck M runs from store 3; therefore, the next stop as determined from Table 7.2.2-1 is store 10, which can achieve a saving of 9.2 km. Similarly, truck S runs from store 2 to store 16, which can achieve a saving of 5.3 km. It should be noted that the load for each truck is now recorded as: L = 14, M = 13, S = 8.
4. Repeat step 2 for the next store in the route. At this stage, it is necessary to note that stores which have been allocated (stores 1, 2, 3, 10, 15, 16) are removed from the next consideration. For truck L, the largest saving from store 15 is store 20. For truck M, the largest saving from store 10 is store 9. For truck S, the largest saving from store 16 is store 19. Thus, the total truck load for the routes is: L = 21, M = 15, S = 14.
5. Repeat step 2 from the outcome of step 3. For truck L, the largest saving from store 20 is store 13 (after removing the allocated store 19). For truck M, the largest saving from store 9 is store 14 (after removing the allocated store 16). For truck S, the largest saving from store 19 is store 18 (after removing the allocated store 9). The total truck load for the routes is: L = 25, M = 23, S = 15.
6. Repeat step 2 from the outcome of step 4. It is noted that trucks S and M may exceed their capacities if further stops are allocated. In fact, it turns out that truck M needs to run one more trip after returning to the hub from store 14 to load another batch of products.

By following the steps previously, the trucks' routes are available as shown in Table 7.2.3-3 and Table 7.2.3-4.

7.2.4 Insertion-based heuristics

Herein, 'insertion-based heuristics' refers to a method that considers the optimal points in different sequences and ensures reasonable total distance by section-by-section testing and comparison. This method inserts optimal points into the temporary route at different positions, then chooses the route with the shortest inserted section as the extend route to ensure that the total travel distance is considerable. The steps in insertion-based heuristics are much more complex than the steps involved in calculating savings; numerous calculations are involved in the solution processes.

Table 7.2.3-2 **Distance between hub to stores and stores to stores**

	0	1	2	3	4	5	6	7	8	9	10	11	12	13	14	15	16	17	18	19	20
0		3.7	2.9	3.4	4.5	12.6	7.6	6.6	9	11.2	14.8	8.5	11.4	11.9	10.1	16.3	12.7	17.7	12.3	15	12.5
1	3.7		2.3	3.6	5.4	10.9	4.5	5.7	10.5	12.1	16	10.9	10.5	9.6	11.6	12.6	10.9	18.2	14.3	16.4	13.7
2	2.9	2.3		2.1	7.6	15.8	7.7	4.4	10.9	9.3	16.6	8.1	9.2	12	9	16.5	10.3	15.4	14.7	13.8	12.8
3	3.4	3.6	2.1		7.1	16.5	8.5	4.4	9	8.7	9	6.5	13.2	12.8	12.4	17.3	9.6	14.9	15.4	11.7	16.5
4	4.5	5.4	7.6	7.1		9.2	10	9.9	9.3	13.9	6.4	12.1	13.6	12.8	16.4	11.2	15.3	21.5	8.1	17.1	16.9
5	12.6	10.9	15.8	16.5	9.2		17.6	20	5.1	15	10.3	13.3	21.9	21.1	21.9	21.1	10.7	26.2	3.5	6.3	25.2
6	7.6	4.5	7.7	8.5	10	17.6		11.9	16.2	34.1	12.5	16.6	4.7	3.8	33.5	8.7	33.6	33.9	18.7	43.6	7.9
7	6.6	5.7	4.4	4.4	9.9	20	11.9		12.6	7.2	8.9	8.5	20.2	18.4	11.3	23.3	8	11.7	19	21.4	22.5
8	9	10.5	10.9	9	9.3	5.1	16.2	12.6		8.5	3.8	6.8	20.7	18.9	11.1	17.4	7.4	17.4	4	5.7	23
9	11.2	12.1	9.3	8.7	13.9	15	34.1	7.2	8.5		5.5	3.7	37.2	35.4	3	10.1	2.2	10.6	11.8	8.9	39.5
10	14.8	16	16.6	9	6.4	10.3	12.5	8.9	3.8	5.5		4.4	17.4	15.6	8.6	43.3	7.9	13.7	7.9	9.9	19.7
11	8.5	10.9	8.1	6.5	12.1	13.3	16.6	8.5	6.8	3.7	4.4		22	20.2	6.1	24	3.1	12.5	8.6	6.7	24.3
12	11.4	10.5	9.2	13.2	13.6	21.9	4.7	20.2	20.7	37.2	17.4	22		3.3	33.7	11.3	33.9	34.1	25	43.9	7.5
13	11.9	9.6	12	12.8	12.8	21.1	3.8	18.4	18.9	35.4	15.6	20.2	3.3		35.1	10.4	35.3	35.5	20	45.2	5.7
14	10.1	11.6	9	12.4	16.4	21.9	33.5	11.3	11.1	3	8.6	6.1	33.7	35.1		24.2	3.2	8.1	12.9	15.1	37.5
15	16.3	12.6	16.5	17.3	11.2	21.1	8.7	23.3	17.4	10.1	43.3	24	11.3	10.4	24.2		26.7	40.5	20	29	10.8
16	12.7	10.9	10.3	9.6	15.3	10.7	33.6	8	7.4	2.2	7.9	3.1	33.9	35.3	3.2	26.7		10.9	10.9	6.4	40.2
17	17.7	18.2	15.4	14.9	21.5	26.2	33.9	11.7	17.4	10.6	13.7	12.5	34.1	35.5	8.1	40.5	10.9		28	20	34.4
18	12.3	14.3	14.7	15.4	8.1	3.5	18.7	19	4	11.8	7.9	8.6	25	20	12.9	20	10.9	28		9.3	24.1
19	15	16.4	13.8	11.7	17.1	6.3	43.6	21.4	5.7	8.9	9.9	6.7	43.9	45.2	15.1	29	6.4	20	9.3		9.3
20	12.5	13.7	12.8	16.5	16.9	25.2	7.9	22.5	23	39.5	19.7	24.3	7.5	5.7	37.5	10.8	40.2	34.4	24.1	9.3	

Table 7.2.3-3 Route information

S	Distance	M	Distance	L	Distance
0		0		0	
2	2.9	3	3.4	1	3.7
16	10.3	10	9	15	12.6
19	6.4	9	5.5	20	10.8
18	9.3	14	3	13	5.7
8	4	0	10.1	12	3.3
0	9	4	4.5	6	4.7
		5	9.2	7	11.9
		11	13.3	17	11.7
	41.9		58		64.4

Table 7.2.3-4 Summary of distances by savings method 1

Trucks	Routes	Demand	Distance
S	(hub, 2, 16, 19, 18, 8, hub)	20	41.9
M	(hub, 3, 10, 9, 11, hub)	25	58
	(hub, 4, 5, 14)	23	
L	(hub, 1, 15, 20, 13, 12, 6, 7, 17)	49	64.4
		117	164.3

We present the same example to illustrate this methodology. Choose the nearest store to the hub location as the next store to go to and insert this store into the temporary route. Then pick up the inserted part and calculate the sum of two distances: from previous point to inserted point, and from inserted point to next point. Then calculate the difference between the sum from the previous step and the distance from previous point to next point. Last, compare the differences in different positions and choose the insertion position with the least difference as the new temporary route.

Steps

1. Choose the closest stores to the hub as the second points in the routes for three trucks.
2. According to the location of these second stops, choose the closest stores from the second points as the third points in the temporary routes for these trucks.
3. According to the locations of these third stops, choose the closest stores from the third points as the fourth points in the temporary routes. In addition, balance the travel distance, demand and truckload when deciding the next optimal stops.
4. Start to insert optimal stores into the temporary routes. Insert a stop into different positions within the temporary routes from the previous step.
5. Calculate the difference of different inserted parts and compare the results in different positions. Using truck S as an example, the temporary route is (0, 2, 7, 0) and the current optimal point is store 9.
   - Insert optimal point 9 into different positions in the temporary route: (0, 9, 2, 7, 0), (0, 2, 9, 7, 0), (0, 2, 7, 9, 0).

- Pick up the inserted parts in every route: (0, 9, 2), (2, 9, 7), (7, 9, 0).
- List the equations to calculate the difference of the inserted parts.

Dhub9 + D92-Dhub2 =	11.2 + 9.3 + 2.9 = 17.6
D29 + D97-D27 =	9.3 + 7.2 + 4.4 = 12.1
D79 + D9hub-D7hub =	7.2 + 11.2 + 6.6 = 11.8

6. Choose the least inserted distance to make a new temporary route. As per the example from the previous step, the third insertion has the least distance; hence, the optimal new temporary route would be (0, 2, 7, 9, 0).
7. Use similar methods (i.e. step 3 to step 6) until all tasks are finished. In addition, based on the option requirement, truck S and truck L are used only once while truck M is used twice. Therefore, care must be taken to arrange limited tasks after truck S and then to stop arranging (Table 7.2.4-1).

Table 7.2.4-1 Calculations for truck M and truck L

Truck M	**Store 16 insert (hub, 3, 11, hub)**				
(0, 16, 3)	Dhub16 + D163-Dhub3	12.7	9.6	3.4	18.9
(3, 16, 11)	D316 + D1611-D311	9.6	3.1	6.5	6.2
(11, 16, 0)	D1116 + D16hub-D11hub	3.1	12.7	8.5	7.3
Truck L	**Store 13 insert (hub, 1, 6, hub)**				
(0, 13, 1)	Dhub13 + D131-Dhub1	11.9	9.6	3.7	17.8
(1, 13, 6)	D113 + D136-D16	9.6	3.8	4.5	8.9
(6, 13, 0)	D613 + 13hub-D6hub	3.8	11.9	7.6	8.1
Inserted parts	Equations	Distance			Results

Table 7.2.4-2 Final arrangement part

Truck M	**Store 14 insert to (hub, 4, 18, 5, hub)**				
(0, 14, 4)	C014 + C144-C04	10.1	16.4	4.5	22
(4, 14, 18)	C414 + C1418-C418	16.4	12.9	8.1	21.2
(18, 14, 5)	C1814 + C145-C185	12.9	21.9	3.5	31.3
(5, 14, 0)	C514 + C140-C50	21.9	10.1	12.6	19.4
Truck L	**Store 17 insert to (hub, 8, 1, 6, 13, 20, 15, 12, hub)**				
(0, 17, 8)	C017 + C178-C08	17.7	17.4	9	26.1
(8, 17, 1)	C817 + C171-C81	17.4	18.2	10.5	25.1
(1, 17, 6)	C117 + C176-C16	18.2	33.9	4.5	47.6
(6, 17, 13)	C617 + C1713-C613	33.9	35.5	3.8	65.6
(13, 17, 20)	C1317 + C1720-C1320	35.5	34.4	5.7	64.2
(20, 17, 15)	C2017 + C1715-C2015	34.4	40.5	10.8	64.1
(15, 17, 12)	C1517 + C1712-C1512	40.5	34.1	11.3	63.3
(12, 17, 0)	C1217 + C170-C120	34.1	17.7	11.4	40.4
Inserted parts	Equations	Distance	Results		

Table 7.2.4-3 Summary of assumption 1 by insertion-based method

Truck	Carload	Upload		Total load	Distance		Total distance	Routes
					1st	2nd		
S	20	20		20	32.9		32.9	(hub, 2, 7, 10, 9, hub)
M	30	29	24	53	33.1	38	71.1	(hub, 3, 19, 16, 11, hub), (hub, 4, 18, 5, 14)
L	50		44	44		80.7	80.7	(hub, 8, 17, 1, 6, 13, 20, 15, 12)
				117			184.7	

There are some brief descriptions of processes of insertion-based method: Table 7.2.4-1 shows the arrangement in the first insertion for truck M and truck L, which in the same stage as the truck S mentioned in step 4. Truck M gets (0, 3, 16, 11, 0) as an optimal temporary route and truck gets (0, 1, 6, 13, 0).

Table 7.2.4-2 is the final arrangement part, truck M and truck L involves the arrangement only, because the truck S only uses once and it has finished tasks already. In this insertion, truck M gets its optimal route (0, 4, 18, 5, 14, 0) for second trip, and truck L gets (0, 8, 17, 1, 6, 13, 20, 15, 12, 0) for its only trip (Table 7.2.4-2).

Summary

Table 7.2.4-3 presents a summary of the insertion-based method. The total distance of this method is 184.7 km for assumption 1.

7.3 Provisioning and inventory analysis

Provisioning and inventory are part of the supply support functions of a support system or part of the supply chain in a production environment. The objective of supply support and supply chain is to have the correct part available, in the right quantity, when and where required in the quantities necessary to support either production or maintenance. In this module we will be looking at provisioning from a maintenance perspective and inventory from both a maintenance and a production operations perspective. The term 'spares' refers to all parts required for maintenance (Nyman and Levitt, 2010).

Provisioning entails identifying and obtaining the initial stock of spares required to support a system and/or its support system. Provisioning or spares assessment should:

- Determine the optimum scale (quantity), range (types of spares) and location of spares for each maintenance level.
- Identify long lead items or items likely to become obsolete so that steps can be taken to ensure sufficient quantities of those items are available to support the system throughout its life.

- Achieve spares procurement concurrent with any production requirements.
- Have spares available before a system becomes operational.
- Initialise the support system.
- Support initial operations until demand-based activity can commence.

Provisioning uses input data from nearly all the integrated logistics support (ILS) disciplines and is one of the final outputs of the logistics support analysis (LSA) process. The process of provisioning results in the generation of a provisioning parts list (PPL). Provisioning activities should occur as late as possible in the design and development phase of a system, but early enough to allow procurement and delivery of spares required to support the first operational system. Figure 5.2 shows the provisioning process.

Identifying the requirements for spares is achieved by either using past experience or statistical projection; both methods, however, use assumptions that do not remain constant throughout the life of a system. Therefore, the goal of determining spares is to be as accurate as possible with initial assessments, and then to add just enough items to allow for a margin of error. The following should be considered when determining the quantities required:

- Additional spares to compensate for repairable items during the maintenance process.
- Additional spares to compensate for the procurement lead times required for item acquisition.
- Additional spares to compensate for the condemnation or scrapping of repairable items. Some items sent to a deeper level of maintenance may not be economically feasible to repair.

7.3.1 Inventory model

Figure 7.3.1-1 presents a graph of a theoretical inventory cycle. Along with economic order-quantity concepts, this cycle commonly appears in literature on inventory control models. The ideal stock line (dotted line) is the perfect situation when the stock is consumed smoothly as planned and the replenished stock arrives in a consistent time (i.e. $t_1 - t_0 = t_4 - t_1$). Procurement costs vary with the number of orders placed. If many orders are placed, high material acquisition costs are expected due to extra processes. On the contrary, placing orders less frequently while maintaining a higher level of inventory will increase inventory-carrying costs. It is common for the replenished quantity to be computed as economic order quantity, which is a balanced action aimed at minimising overall costs. The cycle of consumption, reordering and receiving produces a sawtooth 'ideal stock profile' as shown in Figure 7.3.1-1.

However, in reality, demands are not always constant and quite often the reorder cycle changes with time. The solid line in Figure 7.3.1-1 presents a realistic situation such as when the stock level goes below zero (i.e. out of stock). The negative stock level is recorded if the demand remains and the customer is willing to wait until new stock arrives.

Given the basic requirements for spare/repair parts and consumables, a plan is required within the overall integrated logistics plan to ensure that the appropriate material and parts do not fall into negative stock. In practice, it is not feasible to stock a lot

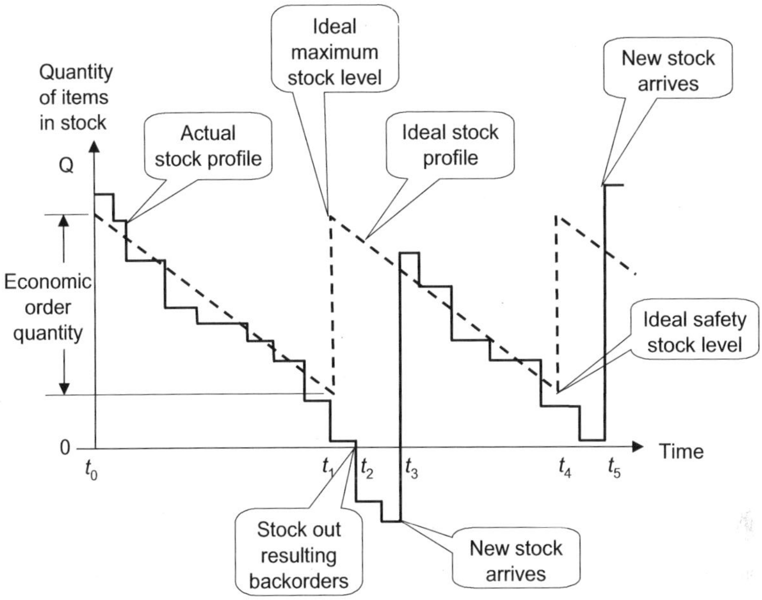

Figure 7.3.1-1 Ideal and actual stock levels

of parts for the entire life of the system because the system's life is very long (in the range of 20~50 years). The stock-carrying cost is therefore enormous, and there is also danger that the components may become obsolete towards the end of the system's service life.

Initial provisioning is based on the estimated maintenance factors provided in the service analysis (i.e. replacement rates, repair and recycle times, pipeline times etc.). Thus, it is important to acquire data to support this type of assessment. We illustrate this type of analysis with a simple example. A communication system has on average 26 component failures per year of the same plug-in element. These elements are replaced from available stock, which is replenished every 2 weeks. The question concerns the level of stock that should be kept to achieve a 90% (or greater) probability of keeping the communication system in service.

To solve this problem, we assume that the failure rate of the elements is constant. The probability of x failures in time t is given by the Poisson distribution:

$$P_x(t) = \frac{(\lambda t)^x e^{-\lambda t}}{x!} \tag{7.3.1-1}$$

The failure rate is 26 components per year or 0.5 failures per week. Hence, the rate for 0 failures in 2 weeks' time is

$$P_0(t) = \frac{(0.5\times 2)^0 e^{-0.5\times 2}}{0!} = e^{-1} = 0.3679 \tag{7.3.1-2}$$

For 1 failure in 2 weeks' time, the rate is

$$P_1(t) = \frac{(0.5\times 2)^1 e^{-0.5\times 2}}{1!} = e^{-1} = 0.3679 \tag{7.3.1-3}$$

Hence the probability of not more than 1 failure in 2 weeks is $0.3679 \times 2 = 0.7358$. In other words, if only one element is kept in stock, it can cover only 73.58% of the time needed to supply the element. The remaining 26.42% is not covered for more than 1 failure. Therefore, it is necessary to investigate a higher stock level.

For 2 failures in time 2 weeks, the rate is

$$P_2(t) = \frac{(0.5\times 2)^2 e^{-0.5\times 2}}{2!} = \frac{e^{-1}}{2} = 0.1840 \tag{7.3.1-4}$$

The additional stock gives the probability of not more than 2 failures in 2 weeks (i.e. $0.7358 + 0.1840 = 0.9197$). Hence if we keep 2 units in stock, we are covered for 91.97% of the time.

If the reliability of the system must be kept higher, say at 99%, we need to keep 4 items in stock. Table 7.3.1-1 shows the probability computed for higher reliability levels.

However, if the failure rate is lower, say only 13 failures per year or 0.25 failures per week, the probability of no more than 1 failure in 2 weeks is

$$\begin{aligned} &P_0(t) + P_1(t) \\ &= \frac{(0.25\times 2)^0 e^{-0.25\times 2}}{0!} + \frac{(0.25\times 2)^1 e^{-0.25\times 2}}{1!} \\ &= e^{-0.5} + 0.5\times e^{-1} = 0.6065 + 0.3033 = 0.9097 \end{aligned} \tag{7.3.1-5}$$

Hence if we keep 1 unit in stock, we are covered for 90.97% of the time. The stock level for 99% certainty can be computed similarly.

The previous example shows estimates derived from predictions or expectations and so may be in error. Estimation errors will of course have a significant impact on the quantity of items in inventory. The provisioning period should be kept at shorter periods to allow for the necessary adjustments based on actual field experience once the system is released to service.

Table 7.3.1-1 System reliability versus spare part stock level

x	$P_x(t)$	Reliability
0	0.3679	0.3679
1	0.3679	0.7358
2	0.1839	0.9197
3	0.0613	0.9810
4	0.0153	0.9963
5	0.0031	0.9994
6	0.0005	0.9999

A realistic period or provisioning period depends on the reliability of the system. In the previous example, the provisioning period can be a few months due to high reliability of the element as well as the short replenishment time. In the provisioning period, enough spare parts are procured to cover the anticipated demand requirements. The provisioning concept should be managed differently for major high–value items and smaller repair parts. In fact, the procurement policies should be flexible to cater to items with different usage rates.

7.3.2 Assessment of contractors reliability

Complex engineering systems require many technical supports from partner companies. In more business-oriented terms, the use of other companies to provide support services is commonly known as 'outsourcing'. The philosophy of outsourcing is to subcontract the service of non-critical parts of a business to external contractors. However, engaging such contractors involves a lot of risks. Good contractors will provide the support services in a way that aligns well with the client's requirements. They are able to do so because their business systems are effective and reliable. Unreliable contractors cannot provide their services at the most-needed times. Assessment of the reliability of contractors relies on historical data obtained elsewhere. The following example illustrates this assessment process.

In this example, a large building project requires the installation of lifts for the building. There are two potential suppliers for the type of lift. To evaluate the suppliers, the performance data of the lifts they manage in other places are listed in Table 7.3.2-1 and Table 7.3.2-2.

We can apply Weibull distribution to model the performance of the contractors. The Weibull distribution is given by:

$$f(t) = \frac{\beta t^{\beta-1}}{\eta^{\beta}} e^{-(\frac{t}{\eta})^{\beta}} \tag{7.3.2-1}$$

Table 7.3.2-1 Failure record of lifts of supplier 1

Failure number	Failed (months)	Failure number	Failed (months)
1	5	11	24
2	8	12	28
3	9	13	30
4	10	14	30
5	13	15	34
6	15	16	35
7	18	17	36
8	20	18	36
9	21	19	37
10	22	20	38

Total number of installations = 200.

Table 7.3.2-2 Failure record of lifts of supplier 2

Failure number	Failed (months)
1	10
2	10
3	12
4	14
5	30
6	35
7	36
8	38
9	45
10	48

Total number of installations = 100.

The reliability is then given by:

$$R(t) = \int_{t}^{\infty} f(t)\,dt = e^{-(\frac{t}{\eta})^{\beta}} \tag{7.3.2-2}$$

Taking the natural logarithm of equation (7.3.2-2), the equation is changed to:

$$\ln\{-\ln[R(t)]\} = \beta.\ln(t) + [-\beta.\ln(\eta)] \tag{7.3.2-3}$$

The slope β and y-intercept $[-\beta.\ln(\eta)]$ can be found by a linear regression method. After inserting the data of the example into equation (7.3.2-3), the constants of the two suppliers can be calculated (Table 7.3.2-3).

We can now compare the reliability of the two suppliers over a number of years (Table 7.3.2-4).

Table 7.3.2-3 Reliability function parameters of suppliers

Supplier number	β	η
1	1.493	175.6
2	1.193	335.4

Table 7.3.2-4 Evaluating suppliers' performance over time

Months	Supplier 1 R(*t*)	Supplier 2 R(*t*)
12	0.9820	0.9813
24	0.9501	0.9578
36	0.9104	0.9378
48	0.8657	0.9063

From the data, supplier 2 is more reliable in the long run. It is important to note that this analysis only takes into account the historical performance data. Care should be taken to also consider other factors such as improvement changes when making decisions on which supplier to choose.

7.4 Remote support technologies

A major characteristic of remote operations-support services is that these activities often occur outside of the vendor company's physical boundaries. In many cases, these services are provided remotely; that is, at a substantial distance from the vendor. Hence a good communication network and commonality of transmission formats are essential requirements for providing remote support services. Communication fosters the ability of the system to provide a good answer to the customer. Artificial intelligence techniques such as expert systems, neural networks and many others have been employed (Karvonen et al., 2002).

7.4.1 Remote diagnostics

In remote diagnostics support, many noncontrollable factors are involved in the process (e.g. changes in the conditions of the machine itself or the input materials). These factors, while normally unpredictable, in many cases are repeatable. However, the major problem with these approaches is that each of these remote diagnostics systems must be created by experts from scratch. This process proves to be very expensive because the situations are one-off in nature. Experience also shows that knowledge gained from one project is specific to that project and so cannot be easily transferred to other projects.

Notwithstanding advances in system monitoring technology there are ongoing challenges for many diagnostics requirements particularly where there are restrictions both on system costs and access to high bandwidth communication channels. The first step to establish condition monitoring for machine reliability assessment is to identify the machine parameters to be monitored. The parameters are selected such that they are easily obtainable from the underlying machine and contain essential performance machine condition information.

The raw data contains a lot of uncorrelated noise and therefore should be filtered. Once the signals are filtered, a mathematical transformation from the time-domain data into a multidimensional data structure. The machine signatures are then constructed from the appropriate cross-sections of the multidimensional data bank. Figure 7.4.1-1 shows the machine signatures of a plasma machine that cuts metal plates by high-energy plasma stream. The signatures indicate different nozzle conditions.

Machine signatures can be identified if unique features can be recognised. In a computerised implementation, the recognition process involves the decomposition of the machine signatures into segments in which features are classified. The features are then matched through a case-based reasoning algorithm that does not aim for an exact match of the pattern but rather tries to find the closest possible signatures in the databases.

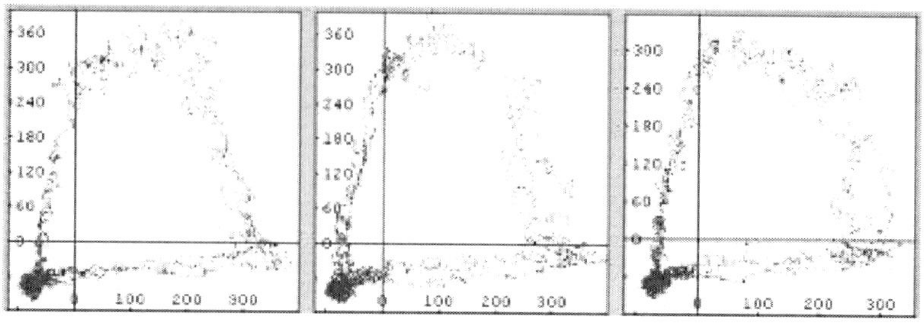

(a) Good nozzles

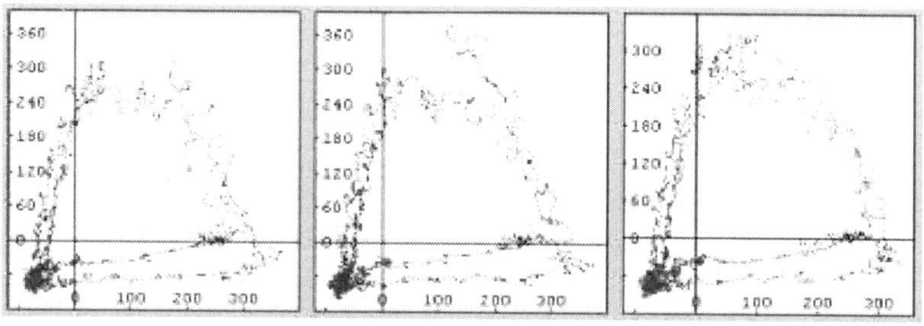

(b) Fair nozzles

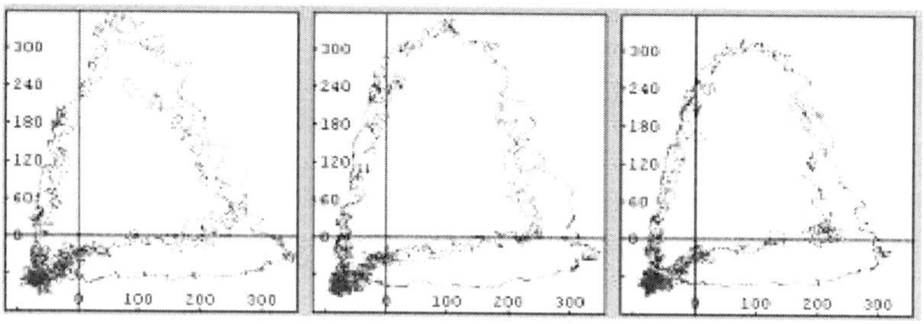

(c) Bad nozzles

Figure 7.4.1-1 Machine signatures created by reconstructed Poincaré sections of voltage signals from a plasma-metal plate-cutting process

7.4.2 Remote expert advice and process optimisation

Expert advice refers to the sup–port service that provide advice to the user on which operating parameters can be used to achieve best results. It involves searching a knowledge base that consists of engineering design knowledge, expert knowledge in scientific investigations, past operating results and rules to evaluate.

There is a need to incorporate new knowledge in addition to field knowledge into the system as the technology develops. Existing systems are generally robust in this sense because they are often designed with a particular problem in mind; however, this intention also renders them deficient in the flexibility to accommodate structural changes of the knowledge representation. Substantial effort is then required to integrate the existing knowledge base with the new knowledge. Changes in knowledge structure also impose problems in the delivery of support to the customer.

7.4.3 Remote training and on-line manual

In order to maintain consistent attention from the trainees and system users, multimedia authoring tools are now available to create applications which can be used for self-training or used as on-line manual. These tools normally run on a PC and so are excellent systems for localised support. However, for customers and users who are far away, other forms of delivery must be employed. For example, graphics information has been successfully displayed by hypertext technology for a local support situation.

References

Aized, T. (Ed.), 2010. Advances in Petri Net Theory and Applications. Sciyo, p. 230, ISBN 978-953-307-108-4.

Graham, K.R., 2009. The life and times of the savings method for vehicle routing problems. ORiON 25 (2), 125–145.

Karvonen, I., van den Berg, R., Bernus, P., Fukuda, Y., Hannus, M., Hartel, I., Vesterager, J. (Eds.), 2002. Part II. Sales and Services, in Global Engineering and Manufacturing in Enterprise Networks, ISBN: 951-48-5739-5, VTT Symposium 224, 9–10 December, Helsinki, pp. 169–276.

Mayer, R.J., Menzel, C.P., Painter, M.K., de Witte, P.S., Blinn, T., Perakath, B., 1995. Information for concurrent engineering (IICE) – IDEF3 process description capture method report, AL-TR-1995-XXXX, September. http://idef.com/pdf/Idef3_fn.pdf.

Nyman, D., Levitt, J., 2010. Maintenance Planning, Coordination and Scheduling, second ed. Industrial Press, p. 314, ISBN: 978-0831134181.

Sterman, J.D., 2000. Business Dynamics: Systems Thinking and Modeling for a Complex World. McGraw-Hill, p. 982, ISBN: 978-0-07-231135-8.

Systems maintenance and reliability 8

Chapter Outline

The concept of reliability is closely related to the risks of malfunctioning of systems. If a system continues to do what it is supposed to do, it is considered reliable; otherwise, it is unreliable. Reliability is associated with the length of time of the correct operation of the system. Thus, reliability is measured or estimated in terms of non-failure over time, under conditions that are specified for the system to work. However, if the system does more than what it is supposed to do, there is no separate concept that expresses the degree of exceeding customer expectations.

A system is designed with the integration of a number of components that work together to produce a certain effect. Reliability can be measured both at system and component levels. System reliability can be traced using knowledge of component reliability, and is therefore estimated according to a variety of concepts. First, system-reliability computation requires knowledge of how the system is put together. The conditions under which the system is supposed to work should be specified; there is no point in estimating the reliability of a system if the conditions are not favourable to the system. Second, the system is expected to fail at some point. Note that failure is not necessarily proof that the system is unreliable but may instead be the imposition of a failure scenario upon the operation. Accordingly, times must be specified for potential failures to be assessed.

By combining a system's components, integration requirements, conditions and time, its reliability can be evaluated according to the satisfaction of the function the system is expected to perform (Niebel, 1994). System reliability is the percentage of non-failures that occur within a unit of time. There are several other measures of reliability, including:

- MTBF, which stands for 'mean time between faults'. This is the expected time between two faults that affect operation of the system (the faults can occur in different ways).
- MTTF, which stands for 'mean time to failures'. This is the expected time between two failures of the whole system, not including human or operator error.

- MTTR, which stands for 'mean time to repairs'. This is the expected time at which a repair must be done before the system fails.
- Failure rate, which is the number of failures that can occur within a unit of time.
- Failure probability, which is the percent of failures that can occur within a unit of time.

Maintenance plans are developed to resume or maintain the asset in operation. Knowledge of system reliability is a key requirement for success.

8.1 Types of maintenance

Maintenance activities can be managed by classifying the types of required maintenance and assigning resources to handle them.

8.1.1 Classification by level of maintenance

Maintenance levels form the baseline for determining which specific maintenance tasks are assigned to each servicing. They are a means by which to select the scope of maintenance and the required skill levels to perform it. Generally speaking, two levels of maintenance can be recognised:

- Operational maintenance.
- Deep maintenance (known as 'major maintenance' in mining and telecommunication industries).

Operational maintenance, which is characterised by on-system maintenance, can be described as 'repair and return to user'. The challenge is to fit the available time for maintenance into the operational schedule. This level of maintenance includes tasks that are regarded as nontechnical. These are the maintenance tasks that may be performed by authorised personnel who are appropriately trained to do them and who do not need to be technical-maintenance personnel. Operational maintenance includes fault identification and preservation actions by supply or distribution personnel where these actions are required to ensure the technical integrity of material. Operator maintenance, which is another subcategory of non-technical maintenance must be performed by authorised personnel who are qualified to operate the material (e.g. specific servicing of aircraft by pilots). Operator maintenance includes routine servicing, parade servicing (military), functional tests, non-technical inspections, authorised-operator repair tasks, and operator tasks during scheduled services.

Operational maintenance generally occurs immediately before or shortly after the engineering system's scheduled operation. It is done by the maintenance and operational crew at the base and sometimes includes necessary minor maintenance procedures. This type of maintenance is done according to original equipment manufacturer (OEM)-designated requirements; the OEM's engineering group provides the maintenance schedule to make sure the aircraft operates and safely meets its through-life capability requirements.

Deep maintenance actions typically involve repair of reparable components off-system, also known as 'repair and return to the supply system'. This type of maintenance is part of the technical maintenance regime that must be performed by personnel who are qualified, competent and authorised in material-specific instructions. It includes technical inspections, testing, calibration and corrective maintenance actions beyond those authorised as non-technical maintenance tasks.

Deep maintenance is much longer in duration and requires more scheduled maintenance hours. For aircraft it usually lasts between 55 and 155 days, or about 1800 hours. Note that the terms 'major maintenance' and 'heavy maintenance' are more commonly used in mining, transport and telecommunication industries. Deep-maintenance durations can often be driven by limited accessibility to engineering data, especially for ageing systems and offshore OEMs.

8.1.2 Classification by planning regime

Maintenance can be categorised into scheduled and unscheduled regimes; in some literature, these are classified as preventive (scheduled) and corrective (unscheduled). Scheduled/preventive maintenance allows an orderly process to take the asset out of service so that disruption to the service to customer can be managed by, for example using an alternative equivalent asset of substituted service or a prior notification. Unscheduled/corrective maintenance is reactive to the process and will be discussed in the next section.

The objectives of operating maintenance schedule are to

- Minimise the number of breakdowns on critical equipment.
- Reduce the loss of production that occurs when equipment failures occur.
- Increase the productive life of all capital equipment.
- Acquire meaningful data relative to the history of all capital equipment so that sound decisions about repair, overhaul, or replacement can be made so as to maximise the return on capital investment.
- Permit better planning and scheduling of required maintenance work.
- Promote the safety and health of the workforce.

Scheduled maintenance is designed to preserve and enhance equipment reliability by replacing worn components before they actually fail. Scheduled maintenance activities include equipment checks, partial or complete overhauls at specified periods, oil changes, lubrication and so on. The maintenance schedule is part of the technical data developed from engineering analysis of the initial design to be used by maintenance. Because management of resources (man hours) and timing (duration) have been predetermined, there is no excuse for not doing scheduled maintenance tasks unless unpredictable events occur.

Due to ongoing expenses that are incurred even when the asset is operating without any problem, scheduled maintenance often gives an impression to the asset owner that it is costly. Although this may be true for some components, one should compare not only the costs but also the long-term benefits and savings associated with scheduled/preventive maintenance. Breakdowns are basically a symptom of excessive wear and tear. In the longer run, maintenance to protect the machine and its components that are subject to excessive wear and tear will result in savings from smoother operations and an expected increase of effective system-service life. Ignoring the means of reducing the chance of breakdowns means that if breakdowns do occur, costs will be incurred for lost production time.

The scope of scheduled maintenance may increase as a result of analysis of the health conditions of the asset. Irrespective of whether maintenance is scheduled or unscheduled, all maintenance activities should include tasks that collect operation and maintenance data that have been accumulated on the asset. These data will be analysed by maintenance engineering analysis; if found to be questionable, they will be referred to engineering for further determination.

It is possible for modification of the asset to be scheduled in a maintenance session for reasons such as obsolescence or change of operating context. Detailed engineering instructions on how such changes should be made will be required.

Unscheduled maintenance is reactive and corrective in nature. It is reactive because the work requirement is not planned beforehand; instead, the maintenance team is forced to react to the event. It is corrective because the reason for convening it is that something has gone wrong with the operation of the asset. A maintenance manager's duty therefore requires 24 hours on call.

Unlike scheduled maintenance, unscheduled maintenance is unpredictable. When unscheduled maintenance is required, it often means taking away the capability of the system without prior notice and thus affecting many other operational issues. The loss due to the disruption of normal operations can be enormous. In fact, unscheduled maintenance is the number one reason for slippage of the production (or operating) schedule of an asset.

Apart from unexpected events affecting the normal operation of the asset and demanding immediate attention, unscheduled maintenance often comes out of inspection. During routine (scheduled) maintenance, undesirable phenomena such as cracks on non-critical structures due to bird strike or oil leakage of lubrication system can trigger a series of maintenance actions to prevent these minor problems to escalate. The trigger can be defined in maintenance manual, or based on history, experience or other references that indicate possible trend. Availability of such reference information is vital to making right decision. Depending on the seriousness of the problem, the normal maintenance schedule can be changed to accommodate additional unscheduled maintenance.

Maintenance management needs to be able to detect problems in the process of managing a maintenance operation. The following symptoms are good indicators of a poorly conceived maintenance program:

- Low machine and equipment utilisation because of unscheduled work stoppages due to breakdowns.
- Long wait times or idle time of operators due to equipment downtime.
- High scrap and rejects due to equipment unreliability.
- Increased cost of equipment repairs because no regular lubrication, inspection, or replacement of worn parts have been recorded.
- Rapid decrease of productive life of capital equipment even when records show that all scheduled maintenance tasks have been done.

8.1.3 Other classifications

It may be necessary to increase unscheduled maintenance as the asset gets older. In addition, the maintenance requirements for an ageing asset become more intense and the interval between servicings may decrease. This need for this type of maintenance, which is sometimes known as 'opportunity' or 'emergent' maintenance, is usually discovered as the asset is disassembled and inspected.

For older assets, opportunity maintenance that is identified while deep maintenance is being conducted may take longer than the work hours assigned to scheduled servicing. In this case, the workload increases significantly. This increase, in turn, causes problems with planning and logistic support because the opportunity maintenance may not only be completely unexpected but may also require advice from the OEM, the development of specific repair schemes, or sourcing for specific parts/components that are not immediately available in the supply chain. To circumvent such problems, one method would be to include the recursive effects of maintenance on-system MTBF as well as maintenance opportunities that are insufficient to prevent system deterioration. The outcome of his method can then be used to construct and update preventive schedules for a complex system. The method is rather complicated and approximate, which highlights the need for fully effective and efficient interfaces between the maintenance function and supply-support function.

Another type of maintenance, the embodiment of engineering modifications, is actually not a genuine maintenance activity although it is usually carried out by maintenance personnel. This activity is usually conducted during deeper maintenance and is considered to be strictly an upgrade task.

Modification of the asset should be done with approval from engineering. If modification has been approved, modification programs are normally inserted into the DM schedule unless there are specific safety requirements that require immediate implementation.

Extended downtime of the system will affect operation. One practice is to carry out phase maintenance, which means dividing deep maintenance into several phases. Phase maintenance requires reconciliation of the maintenance procedure with the operational schedule; downtime will still be significant. Phase maintenance is often employed on, for example, helicopters that are embarked on ships (because it is often not feasible to adhere to maintenance schedules that do not account for the periods of embarkation).

8.2 Maintenance of in-service systems

To handle maintenance while the asset is in service, an integrated, risk-management-based approach is required. The inherent uncertainty around the need for unscheduled maintenance implies that emergency-maintenance plans should be in place and preferably practised so that the maintenance team can react swiftly and appropriately. This approach has been observed in industries such as wind farming, that have high operational uncertainties. The risk-management approach is integrated with the application of a condition-monitoring system that continuously captures system-performance data (including outputs, environmental conditions, equipment failure records etc.) as the basis for computing the risks of incurring unexpected unscheduled-maintenance costs. The proactive maintenance-management strategy can minimise costly unscheduled maintenance and avoid catastrophic equipment failure through early fault detection. By doing so it maximises the operating return of these assets (in the case of wind farms, these would be power outputs, availability and wind-turbine life cycles) and at the same time minimises equipment downtime and insurance claims, all of which has the effect of enhancing the asset's profitability.

Note that by ensuring a rapid and effective response to equipment and performance problems, the proactive risk-management-based approach is equally valuable to the asset owner, financier or customer (of the product generated from the asset, for example power from the wind turbines) during the warranty period as well as the post-warranty period (Kececioglu, 2002).

Before operation As the name implies, this is a specific servicing designed to prepare the equipment for immediate use. In the case of aircraft, preflight maintenance usually entails removing any protective covers, blanks, locks or other devices that protect the system from potential damage during storage, inadvertent operation or infestation from insects or animals. Before-operation tasks may also include checking of consumables such as fuel and fluid levels, oxygen systems and some mission-critical systems operations. These tasks may be done entirely by the ground crew. This servicing is conducted between sequences of aircraft operation and normally consists of replenishment of fuel and fluids used during the previous flight, cleaning as required, and preparation for the next flight.

After operation This servicing is completed once the asset has completed daily or mission operations and consists of replenishments, cleaning and the fitment of in-storage protective devices.

Special servicing This servicing prepares the asset for a specific role or type of operation. If an item on the asset is used for a specific purpose that is rare, then a specific servicing should be done before the asset commences that role.

Configuration change This type of change can happen due to mission requirements. For example, in the military context, and especially for multirole aircraft such as the F/A-18 Hornet, the aircraft can be reconfigured to conduct different mission roles in a single day. This process requires the fitment of specialised equipment and testing it prior to flight. In any case, configuration changes take time and resources and as such are planned and scheduled to minimise the number of changes required.

Replenishments Replenishments can be conducted during any of the above phases and as a servicing in their own right.

Line maintenance/Minor maintenance Sometimes minor unserviceability may need to be rectified during any of the above maintenance activities. Depending on the nature and the level of complexity required to rectify the unserviceability, the maintenance may be conducted *in situ* or the asset is moved offline to conduct the maintenance. Most assets are designed so that minor maintenance may be conducted without shutting down the entire asset's functional system. For example, if a component becomes unserviceable after the asset has been started but prior to making it ready for use, the asset can immediately be put on hold; maintenance personnel can be called in; and the component can be replaced and tested. Thus the operation is continued with minimal disruption.

In service system, maintenance requires extensive support infrastructure including 24/7 availability of qualified engineers, service stations, fault-detection equipment and so on. Finding ways to optimise this support activity is desirable. One possibility is to change the shift schedule to enhance the efficiency of the existing system by spreading the workload more uniformly across shifts. However, this may not be always possible. In any case, maintenance operational changes should be supported by analysed data.

8.3 Assessing an inspect-replace plan

Scheduled maintenance commonly includes the regular replacement of certain critical parts (Jones, 2006). Determination of replacement time depends on several parameters including failure characteristics, operating practice and repair time.

8.3.1 No immediate repair and immediate replacement

If the replacement plan is to replace the part in every inspection but no action is taken in between inspection times, the availability of the system is equivalent to the reliability of the system if replacement is done without loss of time. If the failure rate is a constant λ = 1/MTBF. The availability A is given by:

$$A = R(t) = e^{-\lambda t} \qquad (8.3.1\text{-}1)$$

Figure 8.3.1-1 shows a typical plot for a system with MTBF = 10 units. The system's availability varies as the inspection–replacement time changes. The plot is essentially an exponential function.

However, if the characteristic of the system's failure is a normal distribution $N(\mu, \sigma)$, where μ is the mean working life of the replacement part and σ is the standard deviation of the distribution, the availability A is given by:

$$A = R(t) = 1 - \Pr(t < t_{\text{replacement}}) \qquad (8.3.1\text{-}2)$$

where $t_{\text{replacement}}$ is the replacement interval of the essential part. Figure 8.3.1-2 shows a plot with $\mu = 10$ and $\sigma = 2$.

Figure 8.3.1-2 shows that the availability of the system drops quickly as the replacement time passes the mean life of the component. In order to ensure higher availability, more-frequent replacement is necessary.

Note that mean working life is not the same as MTBF. The relationship between the two parameters can be affected by standard deviation of the normal distribution.

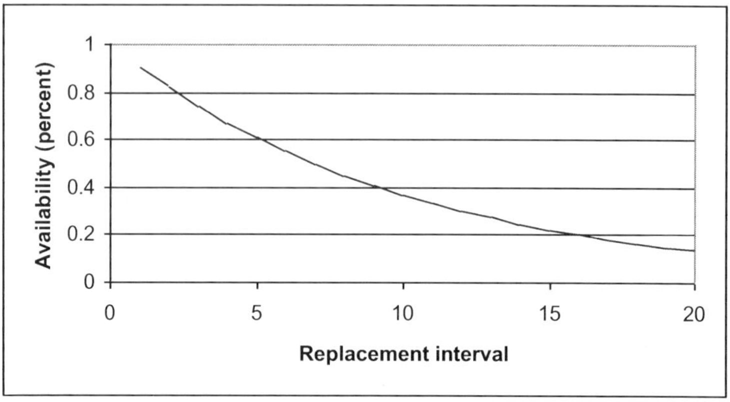

Figure 8.3.1-1 Availability decreases as replacement interval increases based on constant failure rate

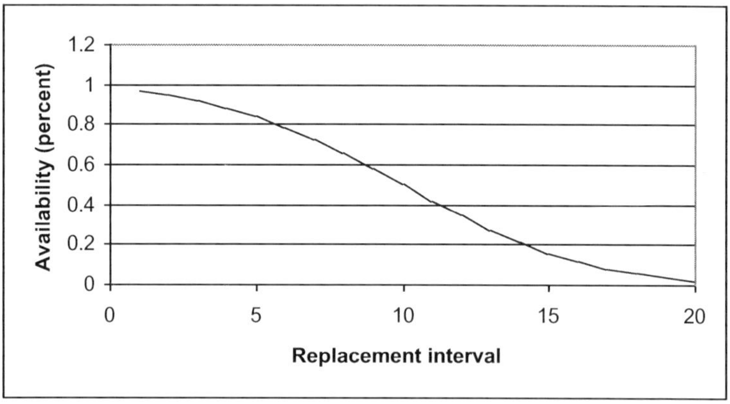

Figure 8.3.1-2 Availability decreases as replacement interval increases based on normally distributed part performance

8.3.2 No immediate repair but replacement takes time

If replacement of the part takes a constant time t_R, the availability is affected by the fact that when replacement is being carried out, the system is not available. The expected uptime is given by:

$$t_{up} = te^{-\lambda t} \qquad (8.3.2\text{-}1)$$

The availability of the system is then given by:

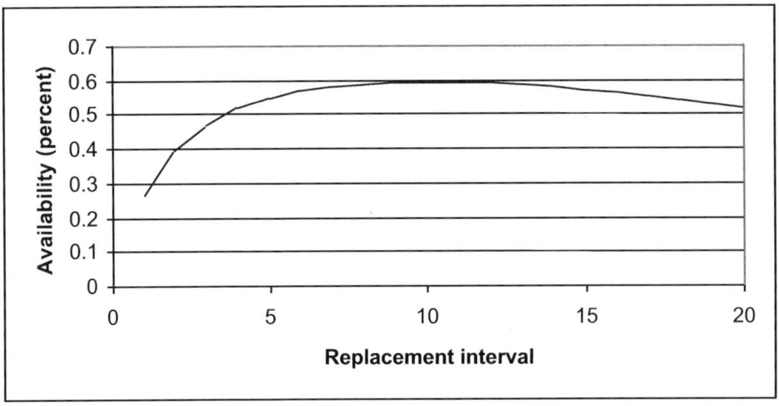

Figure 8.3.2-1 Maximum availability at replacement interval decided to be MTBF

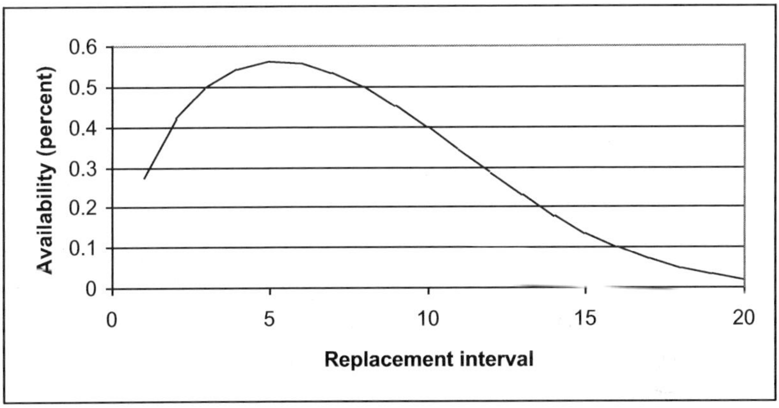

Figure 8.3.2-2 Maximum availability at about half of the useful life of the components replacement interval decided to be MTBF

$$A = \frac{t_{up}}{t_{up} + t_R} \quad (8.3.2\text{-}2)$$

Using the same parameters as in Figure 8.3.1, Figure 8.3.2-1 shows a typical plot with a replacement time $t_R = 2.5$.

Similarly, if the characteristic of the system's failure is a normal distribution $N(\mu, \sigma)$, with $\mu = 10$ and $\sigma = 2$, the expected uptime is given by:

$$t_{up} = t\,\Pr(t \geq t_{replacement}) \quad (8.3.2\text{-}3)$$

The availability can then be computed using Equation 8.3.2-2. Using the numerical example, the availability can be calculated as in Figure 8.3.2-2.

These analyses show that, depending on how much time it takes to replace the part, the availability of a system may be maximised by adjusting the replacement interval.

8.4 Maintenance contract risk assessment

Due to the extremely long-term commitment involved, a services-and-support contract presents a high risk to the service provider's business. If the risks are not well understood, small hiccups during the term of the services-and-support contract will result in sizeable financial losses. Likewise, large mishaps in the operations under the contractual arrangement will impose significant liabilities to the service provider, who may be driven out of business. Hence, apart from setting up the service system, the system designer should ensure continuity of business. It is necessary to estimate the level of risk that the service-virtual enterprise must face over the life of the contract as it provides the service at the agreed-upon price. Both this concept and the technique to enact it are illustrated with a simple worked example (Mo, 2009).

In a typical services-and-support scenario, a service provider (supported by partners within its own supply chain) wishes to bid for a 30-year long-term contract for the services and support of a waste-oil treatment plant. Due to continuous research and development of the oil-treatment process, it is envisaged that the plant will undergo several major upgrades within this 30 years. The service contractor is fully responsible for continual operational availability of the plant to the plant owner.

In order to achieve a high degree of economy for providing services-and-support contracts, a thorough understanding of the nature of the services that will be required is critical to the design of a successful system-support solution. In the aforementioned scenario, the service provider must decide how to assess the viability of service solutions and how to set the fee included in the contract. To analyse this scenario, essential data are solicited from either historical or comparable cases. Here it is assumed that the following data are collected (all costs are in millions of dollars, \$M).

a. There are upgrades required in the 30 years. The year of upgrades t depends on many factors. Table 8.4-1 shows the probability of timing of such upgrades.
b. The cost of upgrade u depends on many factors and can be expressed as probabilities as in Table 8.4-2.
c. The services and maintenance (S&M) costs x in \$M vary over the years based on a probability function in Table 8.4-3. The average S&M costs will increase by \$0.5M every year. So if the figures in Table 8.4-3 are used for determining the S&M costs in Year 1, then the cost figures x will be increased to 20.5, 30.5 and 40.5 respectively in Year 2.
d. After an upgrade, the average S&M costs x in \$M will be reduced by \$1M from the year immediately prior to the upgrade.

The key to risk assessment is to determine a reasonable cost for the contract. This is often computed using an expected value method.

Table 8.4-1 **Probabilities of upgrade timing**

Year t	Upgrade occurring in the year after first install or last upgrade
5	0.4
10	0.3
15	0.2
20	0.1

Table 8.4-2 **Probabilities of upgrade costs**

Cost u (in \$M)	Probability
20.0	0.1
40.0	0.2
60.0	0.2
80.0	0.2
100.0	0.3

Expected year of upgrade: $$\mu_t = E(t) = \sum_{i=1} t_i p_i = 10 \quad (1)$$

Expected upgrade cost (\$M): $$\mu_u = E(u) = \sum_{i=1} u_i p_i = 68.0 \quad (2)$$

Expected S&M costs depend on the time after first operation or the years of upgrade t. Because the expected year of upgrade is 10, there are two upgrades in 30 years (in years 10 and 20). The expected S&M costs at different year y up to the year before an upgrade occurs is given by Equation 3.

$$\mu_{x,10} = E(x,y) = \sum_{i=1} (x_i + 0.5y) p_i \quad \text{where } y < 10 \quad (3)$$

$$\mu_{x,20} = E(x,y) = \sum_{i=1} (x_i + 0.5y - 1.0) p_i \quad \text{where } 10 \le y < 20 \quad (4)$$

Table 8.4-3 **Probabilities of services and maintenance costs**

Cost x (in \$M)	Probability
20.0	0.4
30.0	0.5
40.0	0.1

$$\mu_{x,30} = E(x, y) = \sum_{i=1} (x_i + 0.5y - 2.0) p_i \quad \text{where } 20 \leq y < 30 \tag{5}$$

If the marginal rate of return r is 10%, the net present value of total costs of the service contract in \$M is

$$S = \sum_{i=0}^{9} \frac{\mu_{x,10}(y)}{(1+r)^y} + \frac{\mu_{x,10}}{(1+r)^{10}} + \sum_{i=10}^{19} \frac{\mu_{x,20}(y)}{(1+r)^y} + \frac{\mu_{x,20}}{(1+r)^{20}} + \sum_{i=20}^{30} \frac{\mu_{x,30}(y)}{(1+r)^y} = 319.047 \tag{6}$$

Note that the support system is still servicing in Year 30; however, no further upgrade agreement is in force in the final year of the contract unless a new agreement is made.

Equation 6 provides only an estimate of a likely total cost S of the contract. To determine whether this is a good deal for the asset owner and/or a risky endeavour for the contractor, the distribution of the total cost must be computed.

We set up a simulation model to calculate an instantaneous total cost for several simulated scenarios. We use j to denote a scenario generated by a random number generator that determines the corresponding stochastic values in tables 1–3. Note that j is generated separately for all variables so that their values are independent. The model can be represented by the following equations:

An instance of years of upgrade due to random number

$$j = t_k |_j \tag{5}$$

where k = 0, 1, 2, 3, … (max. 5), and subject to constraint

$$\sum t_k |_j \leq 30 \tag{6}$$

is possible.

An instance of cost of upgrade due to random number

$$j = u_k |_j \tag{7}$$

and an instance of S&M cost due to random number

$$j = x |_j \tag{8}$$

are also possible.

For upgrade period $k \in$ m, the S&M cost of the plant is given by

$$x |_{j,y} = x |_j + 0.5y - 1.0k \tag{9}$$

where y is the year at which the S&M cost is evaluated.

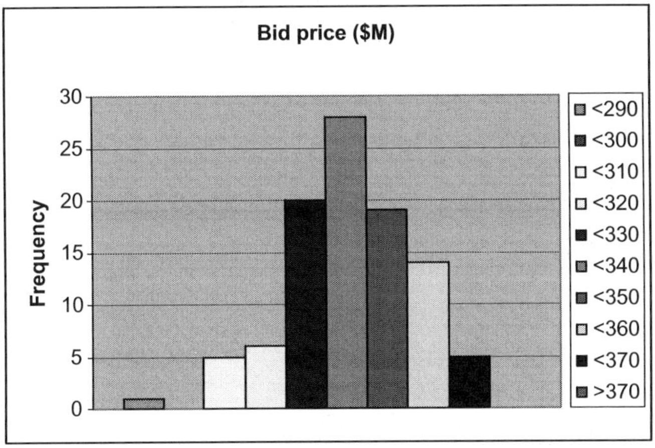

Figure 8.4-1 Frequency of possible total costs for the service contract

Hence, the net present value of the instance of total cost is given by

$$S = \sum_{k=0}^{5} \sum_{y=0}^{t_k l_j - 1} \frac{x \,|_{j,y}}{(1+r)^y} + \sum_{k=0}^{5} \frac{u_k \,|_j}{(1+r)^y} r^i \tag{10}$$

The result of 200 runs of this simulation is shown in Figure 8.4-1.

It is important to analyse the risk of setting a quotation value. If we assume a normal distribution, the mean and standard deviation of Figure 8.4-1 are \$398.72M and \$47.74M respectively. If the contract cost is set at, say, \$450M (in order to compete), the risk of incurring a loss is 14.1%. Thus, this is a high-risk contract and is not acceptable in normal business practice. If the service provider wants to have a 99% certainty of a profit, the contract sum should be increased to \$509.79M.

This case study shows that a system-support engineer will need to develop a mathematical model that represents behaviour of the system over a period of time, to analyse sustainability of the support solution according to prevailing business conditions. The business requirements of service contracts drive a fundamental change in the way services to complex systems are designed. Minimising cost of replacement part will not be the only objective. The main focus is to maintain or improve the performance with changes that are sustainable. Doing so may mean change of parts at a higher rate, but in this case the savings in better performance from the system will pay for the increased component cost. Analytical techniques vary from case to case but the general process of deriving the risk and distribution is common.

References

Jones, J.V., 2006. Integrated Logistic Support Handbook, third ed. McGraw-Hill, ISBN: 0-07-147168-5.

Kececioglu, D.B., 2002. Maintainability, Availability, & Operational Readiness Engineering Handbook vol. 1 DEStech, p. 814.

Mo, J.P.T., 2009. System support engineering: the foundation knowledge for performance based contracting, ICOMS2009, June 15–19, 2009, Sydney, Australia.

Niebel, B.W., 1994. Engineering Maintenance Management, second ed. Marcel Dekker Inc, p. 363, ISBN: 0-8247-9247-5.

Mid-life upgrades

9

Chapter Outline

Due to their substantial business influence in performance contracting, traditional asset ownership and maintenance-service-operation models do not satisfy the expectations of the customer. In a traditional maintenance-oriented arrangement, a contractor is only responsible for the work of keeping the system in operation. There is no consideration of the degradation of capability due to age and technology changes. The resulting capability will deteriorate over time (Figure 9-1).

Nonetheless, a complex system is an expensive asset item that is necessary to serve the customer's need for years. From the customer's point of view, capability should be increased over time to meet need demand and changes in the system's operational environment. To achieve this goal, typical services for the asset include significant mid-life upgrade, whole-system management, integrated logistics support and others. These services are substantially more complex than routine, reliability-based maintenance. They are generally preferred, from the customer's perspective, as expectations that are included in performance-based contracting (Figure 9-2).

9.1 Systems upgradability

To support complex engineering systems through a path of upgrades as technology advances, a total systems-engineering approach that includes knowledge domains

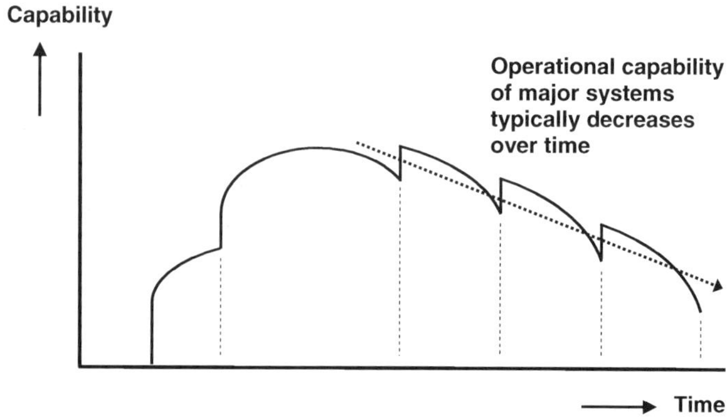

Figure 9-1 Capability degradation through life

such as logistics engineering, supply chain and engineering design should be adopted (Baines et al., 2007). The support team should draw upon knowledge from engineering-intensive organisations and amalgamate with the business-solution aspects of system requirements for the complete life cycle of the system that is being designed and supported (Figure 9.1-1).

In Chapter 2, it was noted that to sustain the operation of an in-service system, it is ideal to maintain the knowledge related to the asset at all phases of the life cycle. Support services that incorporate the ability to upgrade are qualitatively different from the familiar product-based approach in which hard artefacts are delivered to the customer. The upgrade service is a negotiated exchange with the customer, in which the customer is substantially engaged to provide intangible outputs. In this process, system upgradability depends not only on the cooperation of the system user but also the data that are available to support the upgrade project.

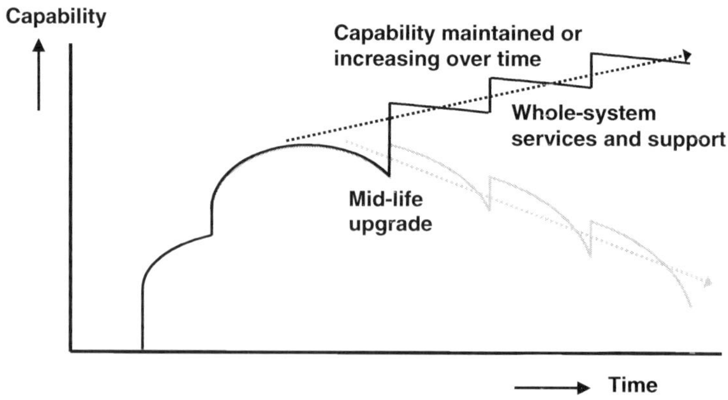

Figure 9-2 Performance-based contracting expectation of improving capability

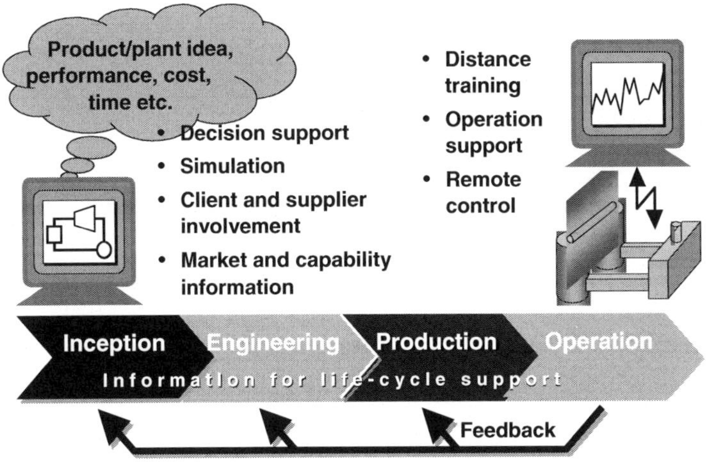

Figure 9.1-1 Complete life-cycle management of complex system

9.2 Data authority

Significant varieties of information must be processed during the service-and-support phase. As the product ages with time and usage, the reliance on data will become pivotal to ensure the original technical integrity is upheld. Or, if a decision is made to deviate from the original baseline information, the delta can be quantified to justify performance, function, safety and environmental impacts. This collation of information will rapidly become extreme as in-service processes are optimised and technology becomes faster. An effective information-management system in such an environment must ensure traceability and the ability to retrieve data in a logical manner both in the immediate and longer-term future (Shen et al., 2004). In a support environment, the data attributes must be established in consideration of longer-term usage by a larger stakeholder group. A wide range of attributes can be applied to an effective information-management system. However, the fundamental attribute groups are concerned with date/time, individuals, information states, information types and classification.

The attributes of data are vital in the ability to retrieve information readily. This need is driven by the continual maintenance and usage of products over time. As a general rule, the more attributes used will enable a refined ability to index, appropriately store, promptly retrieve and logically sort information.

9.2.1 Managing baselines

In very simple terms, a baseline is the configuration of a product at a particular point in the product's lifecycle. The baseline at any time comprises a collection of data and information that can be used to describe the product. It includes such data as specifications, design documents, drawings, software, hardware, test documents and operational manuals. At particular points in time, each element of data is declared at a particular revision status and forms a baseline record. This section will detail only key baselines rather than every single baseline type, and will highlight the challenges of baseline management during in-service management.

To correctly ensure that baselines are declared, recorded and traceable, a software application or database is required because the expected data to describe a complex product are considered significant. This requirement is further justified by the extent of version control that is needed to support the baseline definitions. Once a baseline is declared, the software tool used to capture that data must ensure that the baseline is locked; a master baseline cannot be altered. This process is like taking a photo of the product at a single point in time.

During the in-service phase of a product, a newer baseline will inevitably be required. These updated baselines address in-service changes such as (but not limited to):

- Part replacement to address obsolescence.
- Changes to enhance performance.
- Changes to address stakeholder-changed operational scenarios.
- Potential uses of the product for different or modified intended uses.

The as-built baseline is the initial in-service baseline that is provided to the support organisation. This is the key baseline that will be referred to in efforts to ensure tht the product can be returned to the original build standard. This baseline is not initially intended to change because it defines the delivered product, which has completed testing that has been accepted by the customer as part of the acquisition phase.

Once the product is delivered and used over a period of time, the baseline may change to accommodate repairs, defects, modification, in-service enhancements or warranty issues that in general are not foreseen. In lieu of striking a baseline at specific points in time during development, the baseline of a product is considered to be live at any point in time while the product is in service. After servicing, the product will be delivered against an updated baseline (re-baselined) suite of documentation.

These changes will vary in nature, however, when there is a change across the entire number of one type of product that introduces a change in the function of that type of product. In that case, product data may be formally re-baselined. Smaller changes that are not changing the function are managed on a day-to-day basis and are under the control of a robust configuration management system.

During the in-service phase there may be new non-development artefacts or operational requirements that will result in changes that will subsequently form part of the product description. An example is a specific requirement for a single paint scheme on only one of the products. That product will now have a different defined scheme than other examples of the same product and this difference must be tracked in terms of baseline data-management against each of the products (e.g. by their serial numbers).

Thought must be given to the embodiment of baseness from suppliers and other third parties into the product that is being maintained. Future upgrades and maintainability requirements may require partitions of various baselines. In these scenarios, the configuration-management system must mirror the required partitions so that no development of design will overlap or impact other frozen sub-systems (i.e. sub-systems for which development activities have been completed).

Products that are generated from a manufacturing production line (e.g. trains, planes and automobiles) will require a robust and responsive CMS for baseline management. As the in-service organisation supports the products being produced, the permutation of design changes during the initial and later in-service periods cannot be underestimated. The in-service organisation will experience the challenge of managing the baseline of the products during a period of increased design changes in the infancy and wear-out periods.

Software baselines are subject to the same or (in fact) more-detailed control for the purposes of maintainability and traceability. Software configuration control also extends to the tracking of firmware updates.

Over the life of the product, the in-service organisation must realise that every product may actually be under a different configuration, depending on the status and quantity of each design change or product deviation. Although design configuration captures the part number, the configuration of the supply chain must support the serial number of every part. Thus, this capture is critical to tracing issues that involve batch times and defects of particular parts.

In service it is actually a common practice to cannibalise products in an effort to meet customer-availability requirements. Cannibalisation is the act of removing a serviceable part from one product that has just been obtained, and using that serviceable part to replace a defective part on a product for which delivery is urgently required. The need to rapidly monitor cannibalisation of parts and various modification statuses across a product range is part of monitoring the in-service life of an item. When an item is cannibalised, the newly acquired item will contribute to new product-maintenance intervals. Canalisation is the transfer of parts from a newly inducted product into one that is on a higher priority to deliver. Inevitably there is a risk of one product being used as a cannibalisation source and there is also a risk that when that product is due for release, a robust process for managing its baseline will not be in place. Under these conditions, the configuration may be incorrect.

9.2.2 Validating information

The use of true and accurate data is one of the underlying concepts of ensuring that product designs are technically correct. In some regulatory environments it is critical to validate information before its input into design and support activities. Because information appears in many sources, it is critical to ensure that the data is verified as efforts are made to eliminate all assumptions. Although total elimination is not practically possible, the process of validation will at least highlight any risk of assumptions within the information.

Validation of information may be required at any time; however, in mature systems the information is typically validated shortly after it is received. Given that the information may be in the form of emails, supplier notifications, OEM advice etc.

Table 9.2.2-1 **Methods to validate data**

Completeness	A check that all the expected required data are available. This may confirm that all pages exist, any key references are available and there is no obvious omission of data or key references.
Currency	Sometimes the data provided may be out of date. In these cases it is critical to ensure that the provided data are the latest. On occasion the baseline product may in fact be certified to a particular version or issue to a particular piece of information. Thus the challenge arises to use a more recent design of the legacy-build baseline. Unless the build standard is formally recognised as an acceptable replacement, the original baseline must be complied with.
Applicability	Data may be complete and authoritative, but in fact they may not be applicable. For in-service organisations that receive a large quantity of incoming data every day, a determination of whether the data are in fact applicable should to be made.
Authority/Integrity	The treatment of information will in fact be very dependent on its underlying merit. This merit must be determined where the information is to be used in key in-service engineering decisions. In essence, when information is provided from an authoritative source, the information is to be considered to have sufficient integrity to be a design input. In situations where a desktop review cannot determine validity of data, practical tests or experimentation may be required to validate the information. These methods of validating information arise when the background information is not readily available or information is required for safety-critical systems.

the emphasis is upon performing the validation of data. Making the decision that the data is in fact validated is the responsibility of individuals. Validation of data can comprise methods such as ensuring completeness, currency, applicability and integrity (Table 9.2.2-1).

9.3 Storage and indexing

The importance of a robust indexing system is critical in the retrieval and configuration control of artefacts in the service-support team. Post-acquisition, the day-to-day activities of the in-service organisation will generate a significant quantity of data. Typical data that must be stored for a medium-to-large complex system may vary; however, it ranges from 5 to 500 artefacts per day. Thus in approximately 10 years of through-life support, the quantity of artefacts could exceed 1,000,000. Without robust storage and indexing systems, significant efforts would be required to locate information; if it is not as expected when located, all of that time will have been wasted.

9.3.1 Storage

The storage and indexing systems must ensure that data is adequately stored. Key aspects of these storage systems include:

- Utility for hard-copy data and soft-copy data.
- Verification that applicable security is in place.
- Ready availability, reliability and retention.
- Addressing capacity issues and future support.

The storage systems may comprise many elements to address both soft-copy media and hard-copy media. Storage of soft-copy media is typically a data management system (DMS) that permits users to check in and check out the files for modification. The system will have attributes as discussed earlier, to permit ease of locating. The word 'system' includes such items as filing cabinets and archive arrangements for the storage of other physical media, including paperwork, CDs, DVDs and other types of disks from the vendor. The DMS in this instance will be required to track these physical media. For these media to be considered as controlled, their locations at points in time must be known.

Security should be considered in the establishment of the DMS to ensure that restrictions on the use of stored data are in place. Secure storage of data may vary from a safe (for secret data) to a locked filing cabinet or shelving. For electronic media, consideration must be given to user permissions. Partitioning of a separate repository for various types of classified information may be necessary.

Availability and reliability are emergent issues that typically appear when the DMS is used across a variety of business applications and when the deposited data quantity grows rapidly. Availability issues arise from difficulty in provision of user access, DMS maintenance downtime and unnecessary restrictions on access. Reliability issues arise from the DMS being an unsupported tool (or version of the tool), data corruption and reliability of the underlying infrastructure. Retention of data is critical for a product that is anticipated to undergo numerous upgrades during the contracted service life. Because this long-term retention applies to all assets being stored, an appropriate attribute for the length of storage must be monitored. This is commercially topical because ongoing warranty and support of so many products include contracted periods of data-storage retention. For example, military air-sector design data must be retained until 2 years after the last product of its type has been formally withdrawn from service.

As the in-service organisation and the amount of information that must be stored both grow, the DMS will have to cater to this expanding capacity. Three simple and practical solutions would be to implement auto-archiving, to archive physical media off-site and to expand the on-site storage capacity or compression of data. In any case, a robust DMS will monitor capacity expansion and signal the owner of the DMS to take action when unacceptable issues are predicted.

9.3.2 Indexing

The ability to index data once stored is necessary in order to recall and locate data. Soft-copy media are typically tailored to a particular product. This is a key requirement of any in-service system. Without a properly functioning indexing system it would be virtually impossible to readily locate historic data over the long term. The indexing systems must address:

- Traceability, grouping and linkage to related documents.
- Ability to define groups of data to provoke status reports.
- Ability to relate information to the product breakdown or service element.

Indexing is required to quickly locate any piece of information in the DMS, whether it is hard- or soft-copy data. The index capability is usually structured to search the key attributes that each piece of data can have. In modern DMS, the indexing capability allows relationships to be established between data; these relationships in effect form the traceability of linked information. Because this traceability allows all associated documents that relate to a piece of information to be identified, it enables the impact of changing data to be assessed. In modern DMS tools, simply selecting any linked identifier will retrieve the linked information. The ability to link or associate pieces of information in effect permits groups of information to be established. Typically a tree of information can be built that confers parent/child relationships upon pieces of information. Also, in-service engineers will typically want to see information that is related to specific parts of a product. Hence one valuable grouping to set up is the product breakdown structure in the DMS, and it is also valuable to associate the relevant pieces of information within that structure.

The ability to index and locate specific information permits the reporting and display of relevant information for various in-service stakeholders. This reporting is critical for the status accounting of information and also for the gathering of metrics in order to monitor such aspects as capacity. Because different stakeholders are interested in various pieces of information, reports should be established during the commencement of the in-service phase to minimise the need to rework ever-changing reports. Note that, on occasion, this definition of reporting may in fact identify a set of information attributes that was not foreseen by an engineering cell. In any case, however, it is vital for the DMS to permit flexibility reporting on any tailored set of information attributes. Some systems unfortunately have fixed reports or require additional software to provide reporting capabilities.

9.4 Assessment of information

Assessment of information includes a number of key considerations, such as:

- Relevance.
- Applicability.
- Action determination.

Not all incoming information will be relevant to the in-service engineering (Merlo and Girard, 2004). Thus the very first step in the process of assessing information is to ensure it is actually relevant. Typically a listing of sources of incoming information is defined so that the recipient can easily determine if the information is relevant. If it is not, a determination of relevance must be made. If a new source of information is agreed upon, its details should be included in the list of the information sources.

Applicability is a key determination that should be made by someone with engineering authority. This is needed as the determination of information applicability will then drive a number of consequential actions during follow-up on processing of information.

A professional in-service organisation will capture a decision that information is not applicable, to demonstrate that the information was actually considered and not overlooked.

Once information is determined to be applicable, an assessment is required to identify whether a new action is needed. This action identification and the proposed action are captured in the in-service change management by process or other relevant information stream. In broad terms, the following processing may be to address a defect, address obsolescence, introduce improvements to equipment or publications, or even to implement a contractual change. During the determination of the information it is critical to establish a central processing point in order to reduce the scattershot approach to incoming information.

Should the information-assessment outcome be related to safety or adverse product integrity, an action is generated that addresses the issue. Generating this action must be considered a high priority as delay may have an impact on immediate operators of the product.

9.4.1 Process information

The information will consist of processes that use a set of established protocols that may have separate life cycles for different treatments of data, depending on the determined action to be taken. These protocols may include a formal engineering change, amendments to publications, potential research and development activities, alerting particular stakeholders of issues and investigations.

The processing of information will initiate another life cycle of activities (i.e. action states) that are particular to the process the action requires. These action states should be captured and linked into the DMS. This will provide the required traceability from cradle to grave in this life cycle.

Any of the structured processes should be mapped out; typically the simpler the life cycles are, the easier they will be to manage. One common trap is to initially establish extravagant data flows and many control gates. However, if significant data is entered in a short time, it may be very difficult to reduce the life cycles. Thus processing of data becomes tedious. For this reason, processes should commence with a simple life cycle and minimal control gates. In all flows of information processing, decision points are necessary to permit rejection or halt of the flow. These decision points are required to ensure that stakeholders in the process have the appropriate input at specific points in the decision process and the processing of information. Typically these points are at the commencement, input to content review and conclusion of the process.

It is best practice that the reporting function of the DMS provide reports on the turn-around time (TAT) of particular processes and the outstanding items/actions that require completion. The in-service support-engineering system comprises a number of stakeholders that will be dependent upon some this data and therefore will also need access to or advice on the status of this information.

The value of the ability to prioritise various processing activities cannot be overestimated. With a large quantity of products to support, as well as many stakeholders and in-service emergent issues, the understanding of what is a real priority may actually make or break the overall performance response of the in-service engineering organisation. Many in-service organisations maintain prioritised listings or wallboards to broadcast the true priorities to their teams. Many such lists include a 'drop dead' date that can be traced to the delivery of products or services that are required by the customer.

9.4.2 Incomplete data

The in-service engineering of a product poses challenges for legacy platforms when the service provider changes from company to company or information can no longer be found. Such situations are typical for very long-term projects but not so problematic early in the in-service period. The completeness of information is critical when existing systems must be upgraded, interfaced or repaired. If it is identified that key information is not readily available from the stakeholders or OEM, the filling in or provisioning of such data is required to minimise any input data assumptions. This information generation may include first principles, simple measurements, assessment of similar systems, modelling and organisational knowledge reachback.

9.4.3 First principles

Where a first-principles design approach is deemed acceptable, it should be adopted. A first-principles design approach can be considered as the derivation of design data using analysis from engineering first principles. This derivation can involve hand calculations or computer-based analysis with many types of software engineering tools. First principles alleviate the need to take complex measurements and to create large models. In many software applications the engineering first principles can be automated using inbuilt formulae.

9.4.4 Simple measurements

Simple measurements is an alternative way to obtain missing data if applicable target products can be measured. Such measurements can be taken on the product with all systems turned off, using non-destructive means; they generally include product-component dimensions and other directly measurable characteristics.

9.4.5 Similar systems

Assessment of similar systems enable missing information to be obtained. However, there is a risk of the systems not being exactly the same (i.e. they are similar systems). When an alternative system is identified, the applicable data and information for that system can be obtained and used as information with a (usually) small risk. These assumptions should be captured and verified as part of the typical design-review process.

9.4.6 Modelling

Missing information can also be obtained from the modelling of the product or relevant systems. In some situations, it may be necessary to build physical models (e.g. mock ups of complete product) or computer-generated models (e.g. finite element analysis). The drawback with modelling is that it can become cumbersome and costly. Typically if the missing information is of high importance or accuracy, a model will be generated. The advantage of physical models is that they can be included in their end environment, whereas the software modelling has challenges where the end environmental conditions may need to be modelled themselves.

9.4.7 *Organisational reachback*

The in-service engineering team will build up a network of contacts during the in-support period. This network of agencies and internal organisations will enable reachback to a vast base of knowledge that may include company lessons learnt, identification of like systems, individual experiences or even the identification of missing information. Where information is sourced from third-party organisations, it should be subject to the data review and approval process before use.

9.5 Best time to upgrade

Technology changes all the time. As long as the product supplier maintains a reasonable effort in R&D, it is normal to see incremental improvements throughout the lifetime of the asset. Upgrade involves investment, however, which should be evaluated and analysed to obtain the best outcomes. The method to determine the best time to upgrade can be illustrated with an example.

It is common for a large organisation to install an enterprise system to assist in the management of its operations. The requirement for the enterprise management system depends on the number of transactions that must be processed in a year. The demands of such transactions may vary according to the operating conditions of the organisation (Table 9.5-1).

Each transaction that is not fulfilled in the year will cost $0.00001. The probability of changing operating conditions is estimated in Table 9.5-2.

Table 9.5-1 Number of transactions per year in billions (1 billion = 1,000,000,000)

Years	Demand level (high)	Demand level (medium)	Demand level (low)
1 and 2	600	400	200
3 and 4	800	550	300
5 and 6	1000	700	400

Table 9.5-2 Probabilities of operating conditions

Years	Demand level (high)	Demand level (medium)	Demand level (low)
1 and 2	0.3	0.5	0.2
3 and 4	0.3	0.4	0.3
5 and 6	0.3	0.6	0.1

The existing enterprise management system has been in service for some years. It has capability to handle only 100 billion transactions per year. However, new developments in information and communication technologies and a significant increase in transactions are making it necessary to source a new system to meet the organisation's needs. There are two options for doing so.

Option 1: Buy the most powerful system GX, which can handle 1 trillion (1,000,000,000,000) transactions per year. The cost of this system is $20M. At the end of 6 years, the GX system will be replaced with no residual value.

Option 2: Buy a small-size system SX, which can handle 300 billion (300,000,000,000) transactions per year. The cost of this system is $5M. The system can be upgraded with new CPU, new hardware and new network installations. The cost of upgrades are $1M each. Each upgrade is expected to add 150 billion transactions pa capacity. There is no residual value for SX at the end of its operating life (6 years).

The system can be upgraded every 2 years and management can decide whether to upgrade at those intervals. If an upgrade is not done at the end of first 2 years and an upgrade is requested at the end of fourth year, the cost of the upgrade not done in year 2 will need to be paid as well. The capability after the upgrade will be as good as if the upgrade has been done at the ends of both year 2 and year 4.

Assuming a marginal annual rate of return of 7%, which option would you recommend?

To compare the options, the decision-tree method can be used (Williams, 2011). The process starts by setting up the branch related to the GX scenario (Figure 9.5-1).

From the root node Start at the lower-left corner of Figure 9.5-1, if we decide to invest in the GX option, the investment $5M is represented by B = −5. Having made this decision, there are two possibilities: the market may be high demand or low demand. Whether the market is high or low for the first 2 years, it is possible to have a high or low market for the next 3 years. The scenario investment for SX can be constructed similarly (Figure 9.5-2).

After the tree is constructed, the NPV for each of the branches are computed as shown in Figure 9.5-3.

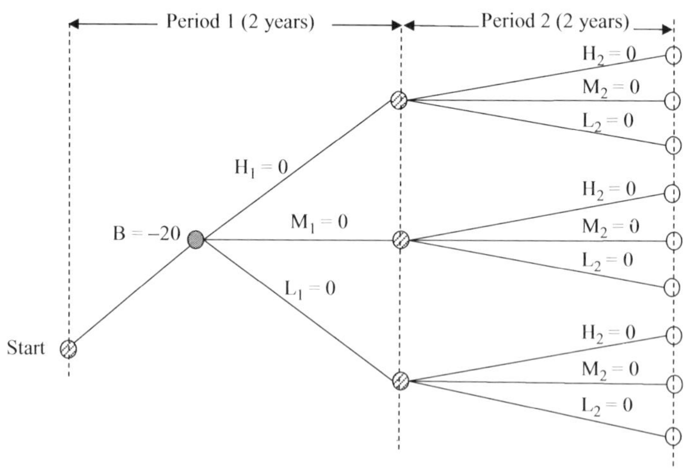

Figure 9.5-1 Setting up decision tree branch for GX scenario

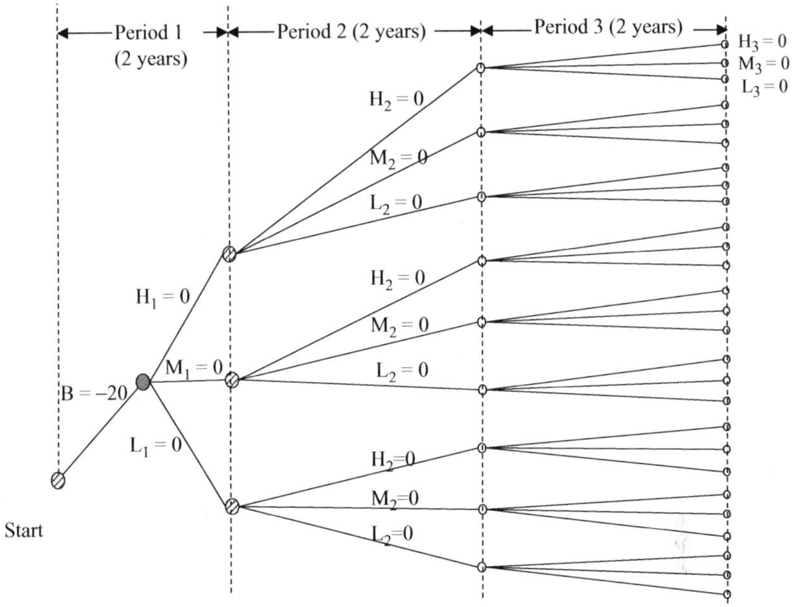

Figure 9.5-2 Expanding all market possibilities due to a GX decision

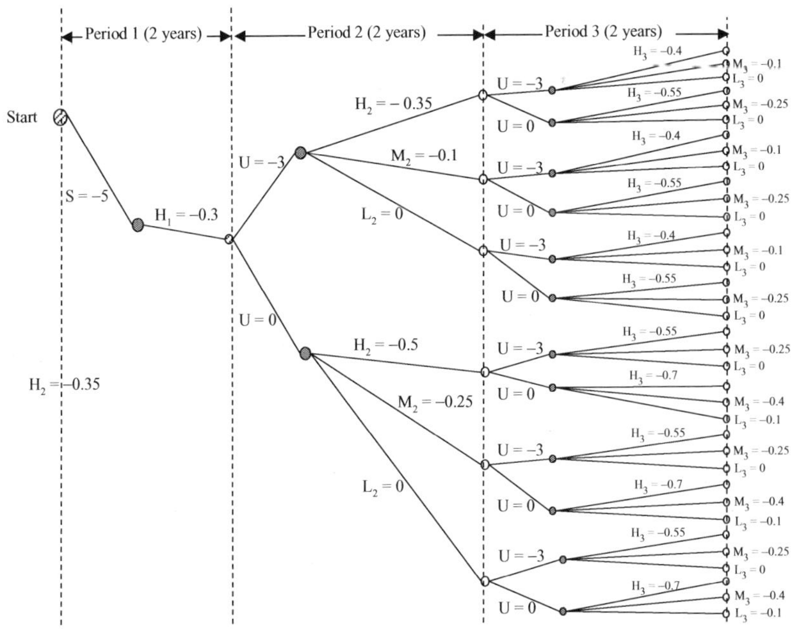

Figure 9.5-3 Setting up decision tree branch for SX scenario

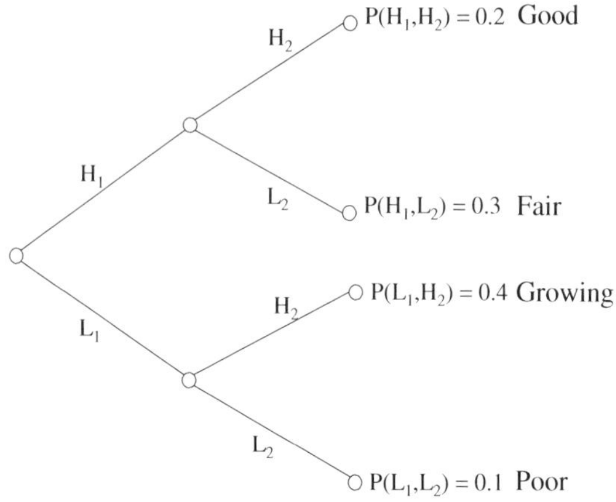

Figure 9.5-4 Displaying probabilities of the end nodes

Next we need to compute the probabilities of each branch. The probabilities and their relations are illustrated in Figure 9.5-4.

The probabilities of the branch can then be computed as shown in Figure 9.5-5.

The probabilities of the top branches can then be computed as shown in Figure 9.5-6.

All other probabilities can then be computed; results are shown in Figure 9.5-7.

Finally, we can compute the NPV under uncertainties for both options (Figure 9.5-8).

Thus the decision can be made at the start by comparing the probability-adjusted NPV for the branches.

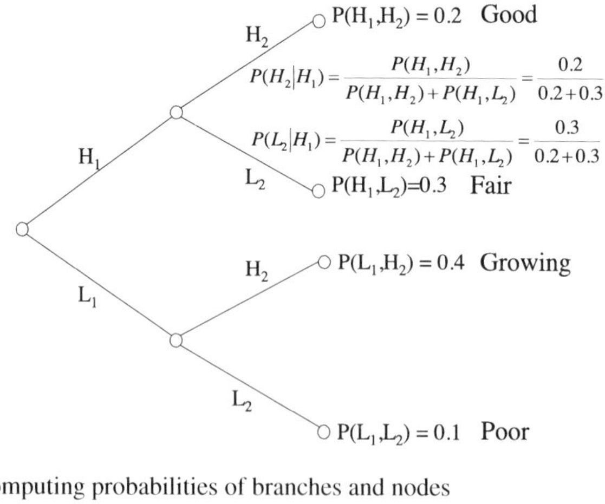

Figure 9.5-5 Computing probabilities of branches and nodes

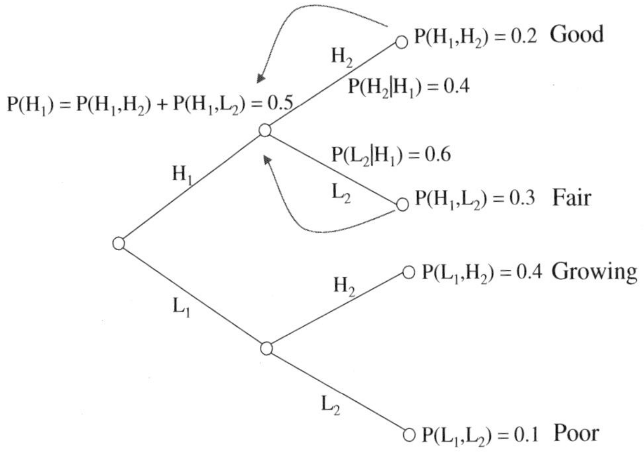

Figure 9.5-6 Computing probabilities of top branches

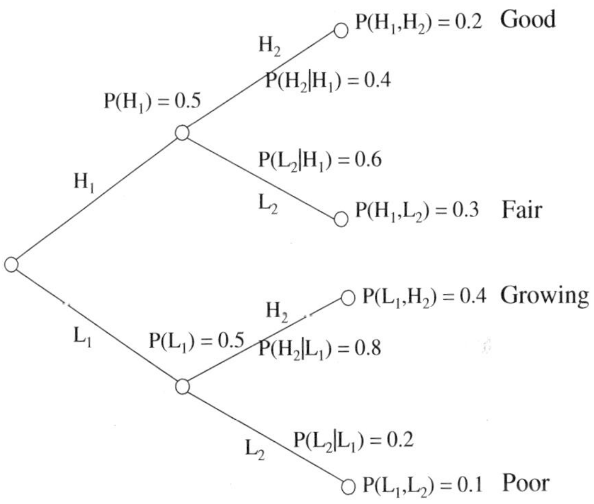

Figure 9.5-7 Probabilities of all branches

9.6 Release to service

Systems undergoing upgrade have gone through significant changes. If not managed carefully, service-release processes can affect product performances, delay activities and impact costs. An engineering change in support solution is a modification to a component of a complex system after that system has entered service. Support engineers should not focus only upon the production and inventory point of view when managing changes. Instead, the work undertaken in support-system design should embrace consideration of the system performance after release to service.

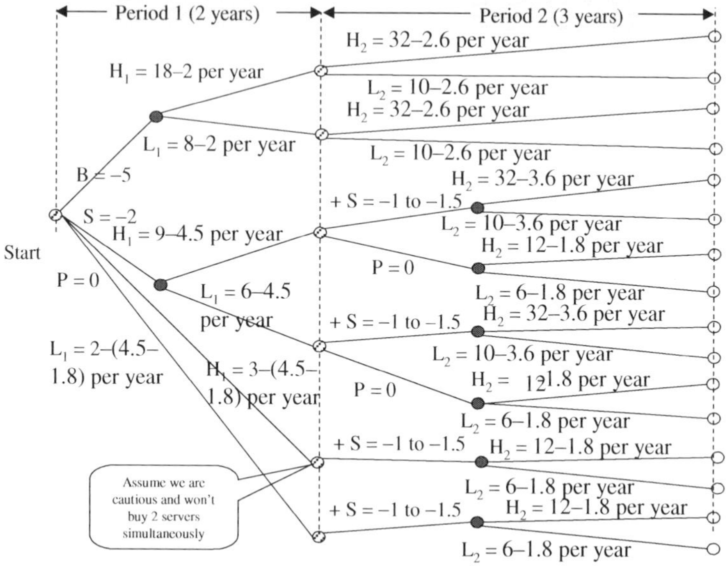

Figure 9.5-8 NPV under uncertainties for both options

Engineering change is an orderly process for evolving the design of a product. Engineering change control refers to the process of controlling how an item is built. This process is controlled by a function that assures that the process of product evolution is done smoothly and with proper authorisation. Engineering change management refers to the process of assigning numbers to the various changes in a product. A good engineering change process provides flexibility without compromising careful control of the manufacturing process. It is centred on the engineering change notice (ECN). An ECN must be entered, approved, implemented and tracked for all changes. ECNs may be implemented immediately or at a future date.

All companies must make changes to existing products, as well as to products under development to meet customer requirements. Engineering changes allow companies to manage these changes efficiently, without interrupting the production of existing products. They allow product and process information such as models, drawings, BOMs and NC programs to be updated and activated according to a user-defined schedule.

Depending upon the complexity of the engineering system being supported, it may be difficult to predict the possible consequences of an engineering change. The immediate impact of engineering is cost and unavailability of the system. However, changes can improve system performance, save money in the long term and enhance system capability. Therefore, all possible consequences should be taken into consideration when initiating an engineering change. Possible impacts include:

- *Cost.* Whatever the change is, it will have an impact on costs. Forecasting the cost of an engineering change is an essential part of the analysis.

- *Availability*. Engineering change can introduce delays in some activities. Affected activities must be re-scheduled and the full impact of re-scheduling upon the business should be assessed.
- *System performances*. Changes can increase system performances due to improvement in process efficiencies. However, future performance after engineering change is not always predictable.
- *Stakeholders*. Suppliers, sub-contractors and work partners or other stakeholders on the system.
- *Other programs*. Effects on other programs upon release of the system to service should be evaluated. The likelihood of effects must be investigated prior to any change study.
- *Life cycle phases*. According to the so-called 'rule of ten', a change detected in the development phase is ten times less expensive than a change detected later. It is fairly rare for a change to produce only one consequence. In an extended enterprise where multiple vendors and contractors are involved in making the engineering change, the initial challenges (particularly for the development of complex products) are to clearly point out the causes that are motivating an engineering change and to evaluate precisely the potential consequences of that change.

References

Baines, T.S., Lightfoot, H.W., Evans, S., Neely, A., Greenough, R., Peppard, J., Roy, R., Shehab, E., Braganza, A., Tiwari, A., Alcock, J.R., Angus, J.P., Bastl, M., Cousens, A., Irving, P., Johnson, M., Kingston, J., Lockett, H., Martinez, V., Michele, P., Tranfield, D., Walton, I.M., Wilson, H., 2007. State-of-the-art in product-service systems. J. Eng. Manuf. 221 (10), 1543–1552.

Merlo, C., Girard, Ph., 2004. Information system modelling for engineering design co-ordination. Comput. Ind. 55 (3), 317–334.

Shen, H., Wall, B., Zaremba, M., Chen, Y., Browne, J., 2004. Integration of business modelling methods for enterprise information system analysis and user requirements gathering. Comput. Ind. 54, 307–323.

Williams, G., 2011. Decision Trees. Data Mining with Rattle and R, Chapter 11. Springer, pp. 205–244, ISBN: 978-1-4419-9889-7.

Architecture for designing support systems

10

Chapter Outline

The shift of complex engineering-products manufacture to service-oriented business environment has triggered businesses to adopt new business models. For example, Internet-based trading such as eBay and online shops have been developed to service customers anywhere in the world. New business models do not just mean new business processes; they can include new products and new types of customers or markets. The concept of product-service system (PSS) was initially developed around the optimisation of sustainability criteria to operations, maintenance and environmental-related issues around the product. It is obvious that there are commercial benefits for companies that move into continuous service-and-support operations of complex products. In addition to the technical requirements of supporting a complex engineering system, a key feature of the new PSS model is the extension of product offerings to service offerings such as maintenance, operations support, training and other after-sales services on a complex product when it is in operation.

However, every complex engineering product and its customer is different; therefore, it is fair to say that each PSS is customised. PSS concept changes contractor roles and responsibilities by shifting the business from product focus to customer focus. Under service-oriented arrangements, the service provider is responsible for the full spectrum of support including ownership, sustainment and operation of assets. Furthermore, contracting arrangements will include incentives and penalties against levels of support-service delivery. The service provider will need to think differently and design output solutions that deliver the desired outputs as well as generate profit under the long-term contract. Therefore, it is important to ensure that each customised service solution design meets service needs and creates new values of use for the customer.

10.1 Characteristics of enterprise architecture for service systems

A PSS contract often involves active interactions, both within the supply chain and with the customer. In a service environment, it is normal that a new, separate enterprise is formed from several independent, collaborating enterprises. There are many risks in this strategy, including collaboration, confidentiality, intellectual property, transfer of goods, conflicts, opportunity loss, product liability and others. To minimise the risks for the new service enterprise, research on enterprise engineering provides an enterprise architecture as a common starting point. The study of enterprise architecture focuses on how enterprises can be designed and operated in an environment when the enterprise missions and objectives are clear. The rationale behind using enterprise-engineering methodologies to guide these steps is to minimise enterprise-design modifications and associated reworkings of the system when the enterprise is put into operation. A support-service system is a dynamic system. Any unplanned change to the enterprise is an impact of uncertainty to enterprise performances. The enterprise architecture (EA) approach provides a structured system to manage services activities such as planning, risk mitigation, standard compliance, operating procedures and controls, and rationalisation facilities.

In order to support decisions on business opportunities by an enterprise architectural approach, the PSS enterprise should have the following characteristics:

- *Use of performance metrics that can be supported by technology*. The support system will be operated in parallel with the complex engineering system. Service is different from the familiar product-based approach in which hard artefacts are delivered to the asset owner. Service is a negotiated exchange with the asset owner (and operator) to provide intangible outputs that are usually coproduced with the asset owner. A service is usually consumed proximately or coincidentally with the time of production. Services cannot be transferred to other asset owners in the same way that products can. Hence the development of appropriate performance metrics is essential and supported by advanced information and computational technologies.
- *Reuse and adaptation of proven concepts*. PSS incorporates system-design knowledge that draws upon principles derived from a wide range of engineering disciplines including systems engineering, logistics engineering, project management, information systems and many others. The knowledge helps the system-support engineer to take into account as many constraints as possible during the system-design phase. These constraints are imposed by the environment in which the complex system and the business are operating. It is therefore important to confirm system design with reference to proven system architecture and experience elsewhere. Ideas and concepts can be reused but care must be taken to ensure that the solution is adapted to the environment in which the solution is applied.
- *Capability for sustaining the support contract*. The performance-based services are characterised by the need to create value for both the asset owner and the service provider. As such, both sides are treated as co-innovators in the design of the service-support solution. Many decisions are made based on incomplete data rather than fully analysed datasets. There are a lot of risks, from the point of view of capability as well as availability over a long period of time. Commitment from both sides is critical to success.

When customers want to outsource a service function, capturing the requirements is the real challenge in order for human intelligence and ability to manage what we know, what other people know and what we don't know. A modelling construct that has comprehensive coverage across all aspects of an enterprise is required.

10.2 Enterprise-modelling methodologies

Enterprise modelling is the underlying science that represents the structure and behaviour of a business entity (Noran, 2005). The enterprise model usually consists of several submodels that have user-oriented metaknowledge structures for communication with human sources of information. The success of the methodology in different areas depends on its representative and explanatory capabilities. These submodels are used to describe the same enterprise from different viewpoints. Due to the complexity of models handled (both in terms of number of components and their relationships existing among components), a modular and incremental modelling approach is often recommended. Most modelling languages use a building-block approach and support the definition of partial models (i.e. libraries of partly assembled building blocks).

Generalised enterprise reference architecture and methodology (GERAM) is part of ISO 15704. The various requirements and components, which all reference architectures must fulfil in order to be considered complete, are synthesised in GERAM's generalised mapping mechanism. Figure 10.2-1 shows the generic enterprise-reference architecture. Based on this architecture, GERAM provides a framework to coordinate a toolbox of

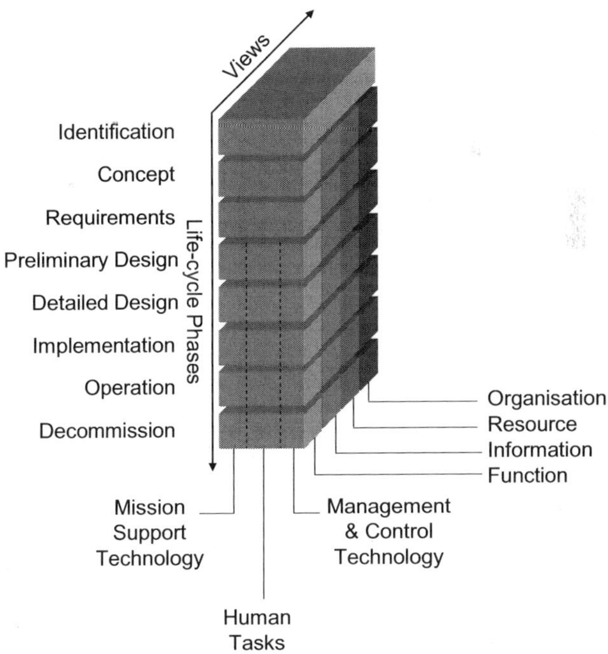

Figure 10.2-1 The generic enterprise reference architecture

concepts for designing and maintaining enterprises during their entire life cycle. It describes the methods, models and tools which are needed to build integrated enterprises. These descriptions are implemented by vendors, mostly in the form of software tools.

In a typical enterprise-architecture tool-set, details of the physical, information and human structures are recorded by modelling formalisms into a database. For example, manufacturing processes are recorded as material flow charts. Management practices are captured as work-flow diagrams; IT system architectures are captured in software engineering tools and so on. This methodology helps enterprise engineers to encapsulate functions and processes within the enterprise.

10.3 Systems architecture

Each enterprise architecture has its particular conceptual foundation that makes it easy use in a particular type of business. For a support system, the product-process-people-environment (3PE) model (Figure 10.3-1) is extended from GERAM to balance the three key elements, viz., technology (i.e. product), management and control (i.e. process), and human (i.e. people).

In the 3PE model, the 3 Ps are:

Product. Application of products with technologies that are efficient and fit-for-purpose and supported by globally recognised industry-technology partners.

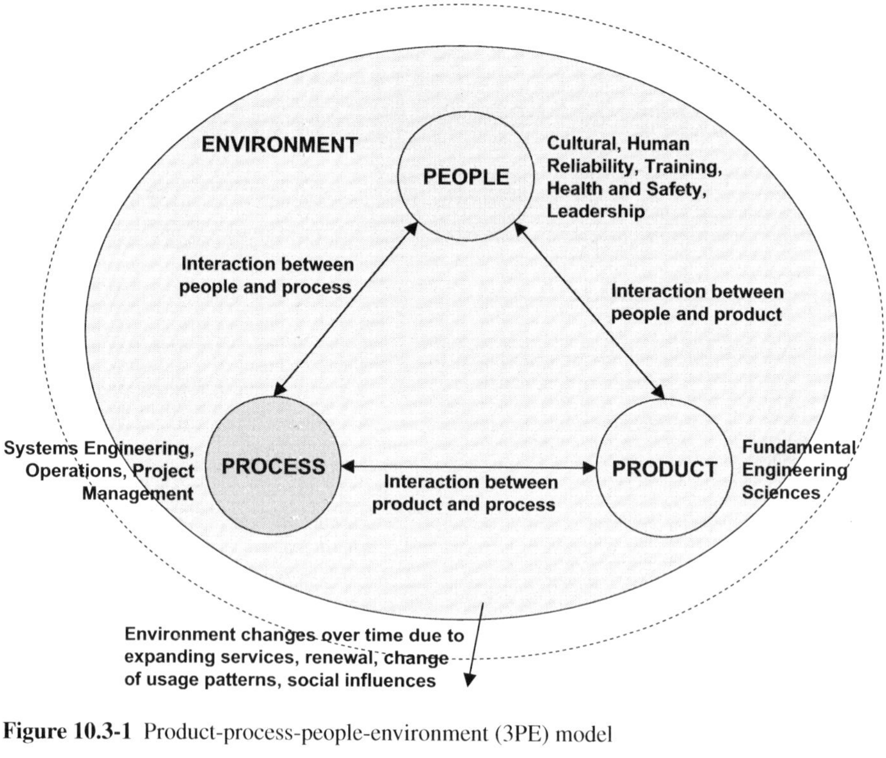

Figure 10.3-1 Product-process-people-environment (3PE) model

Process. Establishment of sustainable systems, good practices and standards that are customised to meet local requirements.

People. Provision of educational training schemes and work opportunities that retain and grow the capability of self-sufficiency in support of the resources and infrastructure investment programs.

In contrast to traditional reference-enterprise-architecture concepts, the 3PE model puts the three Ps in an environment (E) that can affect the service delivery significantly.

Interactions among the Ps are occurring all the time. In the ideal situation, the three elements will adjust to optimise the system's operation to the constraints. However, the environment is always changing. This drives changes of the system or, more specifically, the Ps.

10.4 Support systems design case studies

To understand the essence of the 3PE enterprise architecture model, three case studies with emphases in different elements are discussed in this section. These case studies illustrate how a support system can be designed or adapted to meet the customer's requirements as well as to satisfy environmental constraints.

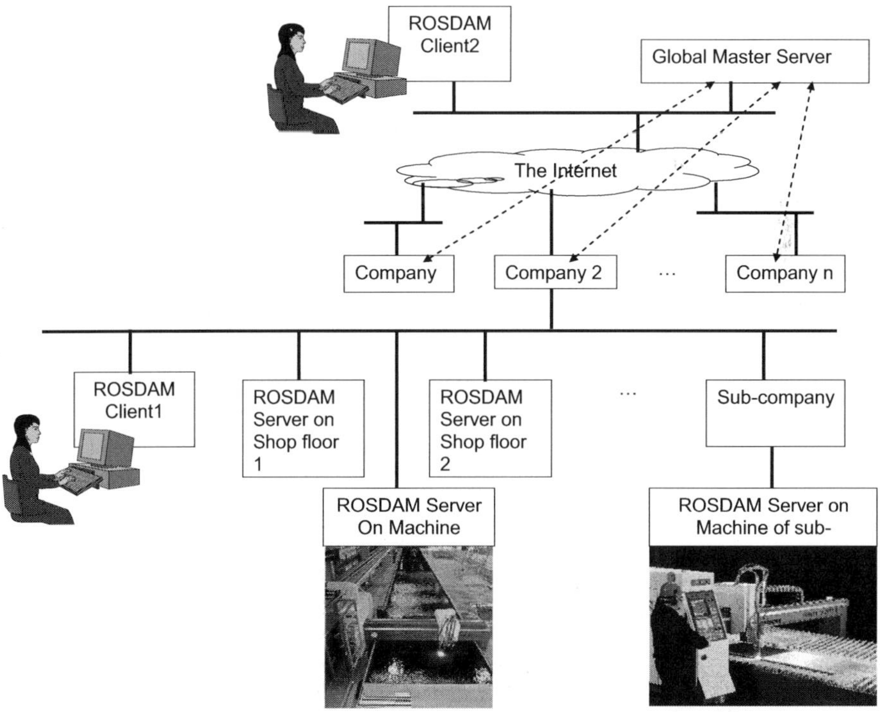

Figure 10.4.1-1 Signal-based condition monitoring service system network structure

10.4.1 Case Study 1: Signal-based condition-monitoring system

System-health monitoring plays a critical role in preventative maintenance and product-quality control of modern complex engineering products. The effectiveness of management can directly impact their efficiency and cost-effectiveness. A condition-monitoring system monitors the products using various classic methods of signal analysis such as spectrum or state-space analyses (Yang et al., 2003). Maintenance decisions are then made according to the prediction of system performance.

Using time-based signals available from normal machine-sensing mechanisms, a CNC machine manufacturer in Australia developed a remote condition-monitoring system for CNC plasma- cutting systems with the aim of servicing the customer anywhere in the world via the Internet. Figure 10.4.1-1 shows the network structure of the system known as ROSDAM. All ROSDAM- enabled machines were configured as servers that had functionality and were communicating with the global master server. Information about the operation of the machines was captured through individual

Table 10.4.1-1 Mapping of service elements in case study 1 to 3PE

Element	Description	Mapped to 3PE
On-machine signal-based diagnostics capability	A new diagnostics software module based on chaotic theory and digital signal processing was developed to assist identification of faults.	Product
Communication networks and IT systems based on client–server model	The controller of the machine was significantly changed from a normal standalone operating system to one that can act as a server in a network environment.	Product–People
Knowledge sharing that transforms customer data to information to knowledge	New data processing algorithms were developed as software modules that were required to process data on machine to knowledge useful to enhance operational efficiency.	People
Engineering information that is integrated for supporting more-effective customer service	Engineering information such as bill of material, machine- configuration management, parts inventory and resources planning were integrated from different sources including CAD, MRP and various manufacturing sources to create a seamless operation database for the machine.	Process
The new system design requires upgrade of field products.	Field upgrade for machines that were already installed at customer's location was progressively rolled out according to contracted maintenance schedules.	Product–Environment

companies' databases. The significantly improved sources of information enabled the product manufacturer to decide the best option that supported operation and maintenance of the plasma-cutting machine from a distance.

In this case study, new elements were required to be developed and integrated with the product (i.e. the CNC plasma-cutting machine). These elements are mapped to the 3PE model as shown in Table 10.4.1-1.

10.4.2 Case Study 2: Global operation-support services

Complex assets are normally built from a large number of components and involve a large number of engineers and contractors (Kamio et al., 2002). In the past, customers as plant owners usually maintained their own service department. However, the increasing complexity of the plant and operating conditions such as environmental considerations require service personnel to have higher levels of analysis and judgment capability.

In managing the design and manufacture of a chemical plant for their customer, a service virtual enterprise (SVE) with several partner companies around the world provided after-sales services to a customer (Figure 10.4.2-1).

Each partner in Figure 10.4.2-1 was an independent entity that was equipped with its own unique capabilities and competencies, and assumed responsibility to perform the allocated work. The support virtual enterprise was designed as a hosting service with a broad range of services including plant monitoring, preventive maintenance,

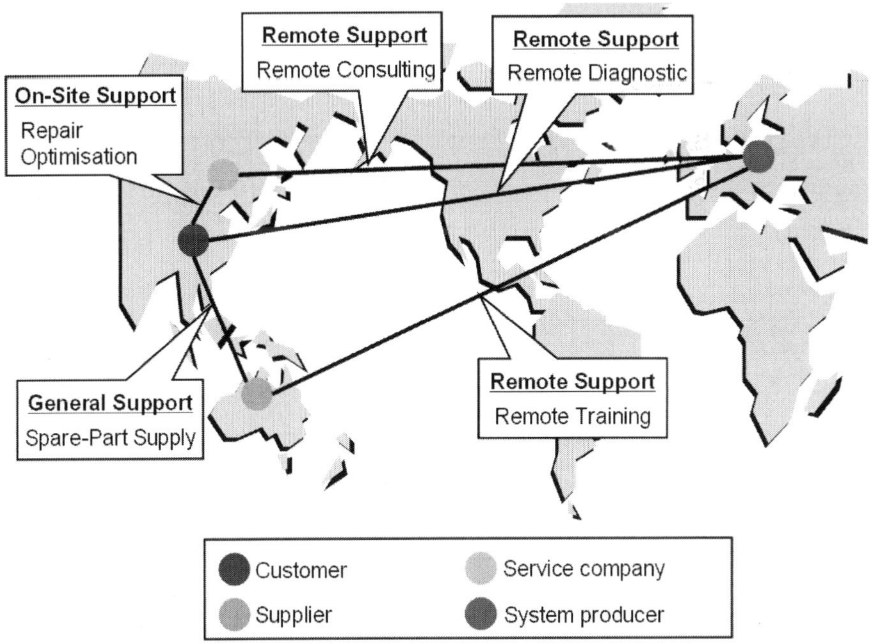

Figure 10.4.2-1 A globally service virtual enterprise

trouble-shooting, performance simulation and evaluation, operator training, knowledge management and risk assessment. Participants in this virtual enterprise had well-defined roles and responsibilities.

Two essential elements in the design of a service enterprise is to develop efficient system architecture and to provide the right resources to the right service tasks. By synchronising organisational activities, sharing information and reciprocating one another's technologies and tools, each partner in the service enterprise will be able to provide services that would have been impossible by individual effort. The support solution therefore requires properly designed systems to support the use of technology in the provision of support services to customers.

In this case study, in order to operate the SVE, new service elements were developed. They are listed and mapped to 3PE in Table 10.4.2-1.

It should be noted that the engineering product remained the same as it had been designed initially. No noticeable engineering change was required on the product itself in order to implement the support service offered by SVE.

10.4.3 Case Study 3: Ship-service system

The ANZAC Ship Alliance (ASA) could be thought of as a virtual company with shareholders comprising the Australian government and two commercial companies, one of which was the ship builder (Mo, 2009). The primary goal was to create 'best value for money by managing all change and upgrades to the ANZAC Ships'.

Table 10.4.2-1 Mapping of service elements in SVE to 3PE

Element	Description	Mapped to 3PE
Inter-company communication networks and IT systems.	New IT and communication systems were installed to enable inter-company exchanged of information as well as personal interactions.	Process–People
Work items synchronised within the project across companies.	Work items were analysed individually so that the link from individual level to group level could be streamlined to ensure minimum duplication of work and conflicts.	Process
Change of human organisation role.	The relationship within an SVE was definitely different from a totally authoritative company structure. A much more flexible human- organisation structure was established.	People
Global access by customer.	The SVE was implemented on the Internet. At the time of the development of this PSS, the Internet was still not entirely effective in some parts of the world, and this environmental constraint imposed significant challenges to the SVE development.	Process–Environment

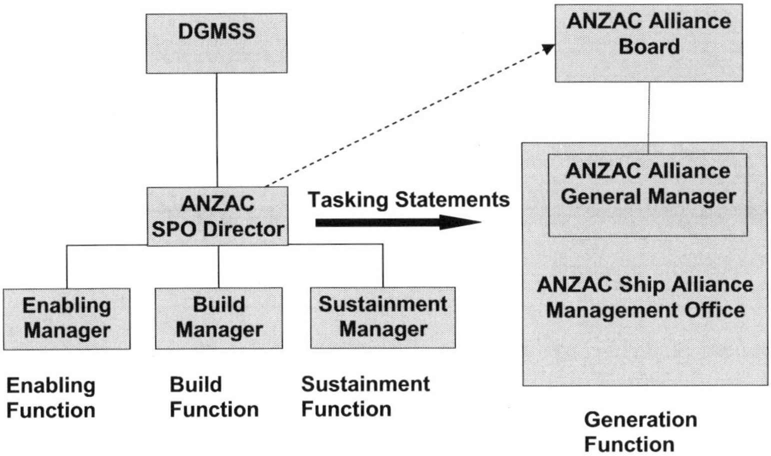

Figure 10.4.3-1 Organisation structure of ANZAC Ship Alliance

Over the lifetime of the asset (30 years), changes due to new technologies, people and defence requirements are inevitable. The organisation structure of ASA can be described as shown in Figure 10.4.3-1.

In this case study, the enterprise was not set up as a legal entity. There was no formal binding agreement among the partners in ASA. In the language of virtual enterprise, the partners were loosely linked organisations such that everything done in the ASA was based on trust. The new service elements that were developed on this premise are listed in Table 10.4.3-1.

Table 10.4.3-1 **Mapping of service elements in ASA to 3PE**

Element	Description	Mapped to 3PE
New procedures and legal processes.	All parties joined the ASA with the understanding of a set of business rules: All parties win or all parties lose; Collective responsibility, equitable sharing of risks and rewards; All decisions to be based on 'best-for-project' philosophy; Access assured to resources, skills and expertise of all parties; All financial transactions to be fully open-book; and Encouragement of innovative thinking and outstanding outcomes. These nontraditional business rules imposed challenges upon the ASA as the service system of the ships.	Process

(Continued)

Table 10.4.3-1 Mapping of service elements in ASA to 3PE *(cont.)*

Element	Description	Mapped to 3PE
Ownership of product-related services rested with ASA.	The enterprise was a joint development of several companies; hence, the ownership of the product-related services had to be resolved. This issue was eventually settled but as all participants were in the defence-business environment, and the customer was a partner of ASA, the company structure at the right-hand side of Figure 5 was able to provide sufficient and reliable background understanding.	Process–Product
Secondment of staff from the three partner organisations.	Because the company status was eventually accepted, the need to develop a set of processes that was acceptable to all staff (who were all seconded from the partner organisations) became urgent. A lot of time was spent in synchronising practices and culture that originated from individual companies. There was confusion in the first year among the staff of the participating organisation about the nature of ASA. This issue was resolved through a number of ASA workshops.	People–Environment

10.5 Transition management

As Figure 10.3-1 indicates, the environment is changing and thus the support system is forced to change as well. Change requires transition planning to minimise the impact upon the bottom line of the support system (Chen et al., 2008). Under the 3PE modelling framework, product change is a matter of investment, whereas process change is management-related. With appropriate practices, it is not an issue. Transition affects people, and it is very often the transition that people resist rather than the change itself. Staff at different stages of system change will have emotional responses that should be recognised. In addition to technical factors, complex organisational, managerial and social factors influence the success or failure of any change project.

Organisations that are transitioning to attaining sustainable competitive advantage need the capacity to generate, renew and use knowledge. In technologically complex industries, the tacit knowledge residing within staff is an intangible asset that is difficult to capture. In the new economy, it has become the top priority for leadership to manage the knowledge in its workforce if the organisation is undergoing transition.

Managing knowledge transfer during transition management can't be effective without good planning. Transition of organisational capability can be achieved by establishing internal structures and processes that aid the creation of competence and

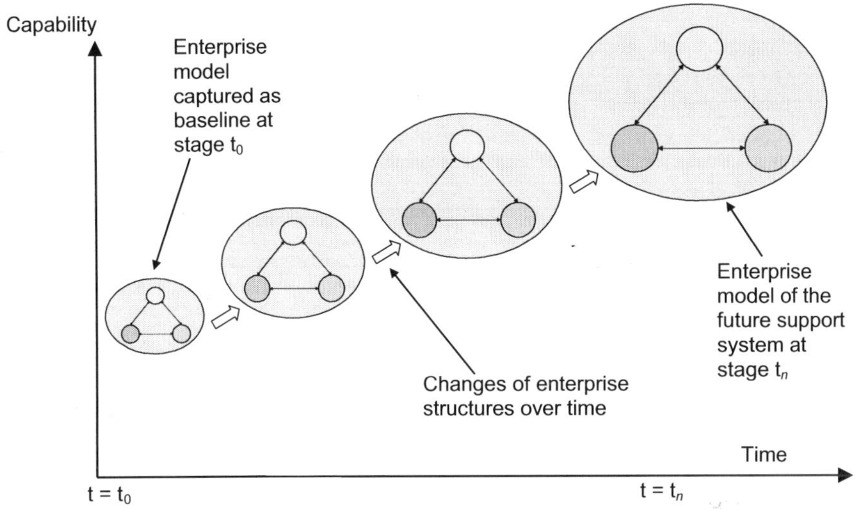

Figure 10.5-1 Enterprise models in a transition trajectory

hence the ability to transform. An enterprise architecture defines methods and tools which are needed to identify and carry out change.

An enterprise model can describe a baseline support system at time t_0 (Figure 10.5-1). The baseline description is practically a snapshot record of the enterprise. The enterprise model enables interpolation between virtual snapshots that lead to the identification of trends and changes in the enterprise architecture. By carefully analysing the evolution and interlinks among different functions, data and processes, a development continuum can be mapped out to form a trajectory (Figure 10.4.1-1). The new (future) architecture covers the additional changing aspect of service system by integrating the concepts of product, process and people to changes in the environment over time.

The enterprise-architecture-design approach focuses on designing the enterprise at different anticipated development stages using established enterprise-design guiding principles. The success of the new (future) enterprise depends on the correctness of the enterprise vision and upon well-managed implementation. Although simulation techniques can assist in clarifying the effect of some design factors, there are many other aspects of the enterprise that can be simulated or foreseen. The transformation is therefore risky and nonresponsive to external environment. Transition management means being able to visualise and predict the trajectory of enterprise changes so that appropriate actions can be taken to maximise benefits in the change.

References

Chen, D., Doumeingts, G., Vernadat, F., 2008. Architectures for enterprise integration and interoperability- Past, present and future. Comput. Ind. 59 (7), 647–659.

Kamio, Y., Kasai, F., Kimura, T., Fukuda, Y., Hartel, I., Zhou, M., et al., 2002. Providing Remote Plant Maintenance Support through a Service Virtual Enterprise. VTT Symposium 224, Global Engineering and Manufacturing in Enterprise Networks. 9–10 December, Helsinki, Finland, pp. 195–206.

Mo, J.P.T., 2009. The use of GERAM for design of a virtual enterprise for a ship maintenance consortium. In: Saha, P., Handbook of Enterprise Systems Architecture in Practice. Information Science Reference (IGI Global), pp. 351–365, ISBN 978-1-59904-189-6.

Noran, O., 2005. A systematic evaluation of the C4ISR AF using ISO15704 Annex A (GERAM). Comput. Ind. 56 (5), 407–427.

Yang, S.Y.S., Kearney, T., Mo, J.P.T., Boland, P., 2003. Machine GP: a new concept in signal-based machine condition monitoring and diagnostics. New Eng. J. 7 (2), 19–23.

Performance and risks analysis 11

Chapter Outline

The recent trend around the world among the owners of complex engineering systems such as aircraft or oil refineries is to include consideration of the sustainment of the system at the very early stages of system development. In general, the asset acquisition project is considered a continuum of four phases, which can be generalised as a capability-systems life cycle (Figure 11-1). The ultimate goal is to attain desired capability levels that can be measured as performance outcomes of systems in-service.

There are two different contracting regimes in Figure 11-1.

- System acquisition agreements, including functional and performance specification of the final system (i.e. the tendering and contracting activities in the acquisition phase).
- Sustainment agreements that specify outcomes and performance requirements for in-service support (i.e. the in-service support contracts in the sustainment phase).

From the industry's point of view, this shift in the defence-acquisition process means longer, more assured revenue streams based on long-term support and ongoing development instead of a series of big 'must win' procurements.

Nonetheless, sustainment agreements often cover a period as long as the service life of the asset. The greatest risk factor in any contract is potential change over time, as discussed in the 3PE model. Understanding of the relationships among the different influencing factors in a support system is necessary in order to minimise and control the risks.

11.1 Value and performance in service

In a performance-based contracting environment, a contractor offering systems-support services needs to design an operations and support system that is sustainable and fit for the purpose, and also needs to demonstrate its value for the money. The advantage of performance-based contracting is the sharing of benefits for both sides of the business. Efficiency gains are shared between the contractor and the owner of the business (Defence Materiel Organisation, 2007).

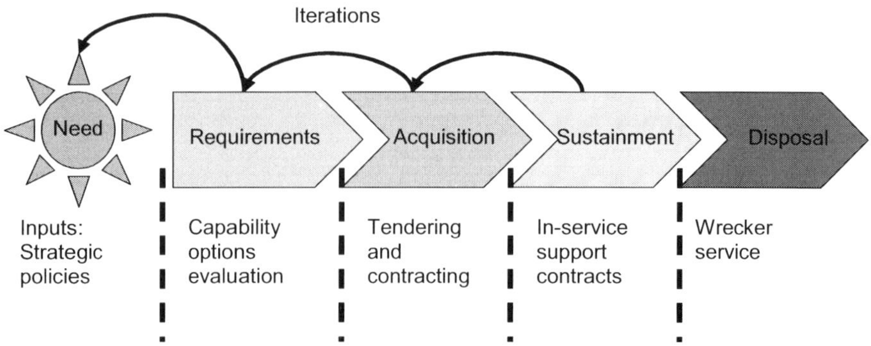

Figure 11-1 Capability-systems life cycle

Depending on the complexity introduced by management, contract payment terms can be described as a function of performance:

$$\{\mathbf{V}\} = f(\{\mathbf{P}\})$$

where $\{\mathbf{V}\}$ is the vector of values perceived by the customer (asset owner) and $\{\mathbf{P}\}$ is the vector of parameters representing the outcome of the contract.

In its simplest form, $\{\mathbf{V}\}$ and $\{\mathbf{P}\}$ can be represented by single value functions such that

V = financial value received on the contract, and

P = an aggregated performance indicator

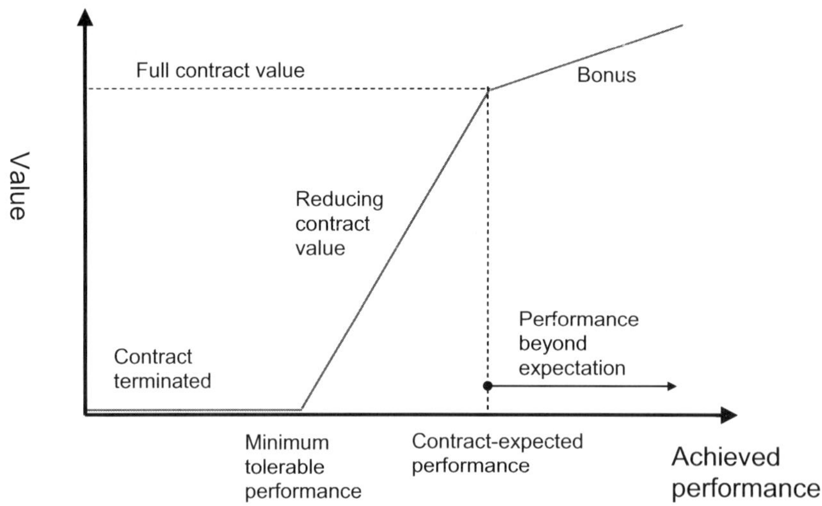

Figure 11.1-1 A typical performance-based payment scheme

A manufacturer of a complex engineering system entering into this kind of contract takes a lot of risks with the contract. A typical contracting framework contains several elements of incentives and penalties; it can be diagrammatically represented (Figure 11.1-1).

As shown in Figure 11.1-1, the contractor is expected to deliver the 'contract expected performance' and be paid the full contract value. If the contractor achieves performance that is better than the contract expected performance, a bonus will be paid in proportion to some kind of performance measure. If the contractor underperforms yet the services provided are tolerable, the contractor will be paid a reduced value proportional to the performance measure. Acceptance of performance should be considered as a whole, however. 'Minimum tolerable performance' refers to performance that is negligible in value because the required outcomes cannot be obtained.

Thus a service contractor must ensure the required performance level of the product in order to receive full contractual payment. Anything below the required performance level will result in reduction of revenue. The rate of revenue reduction can be any linear or non-linear function. Note that if the contract is terminated, the customer may seek compensation from the contractor on damages to the normal business because the system does not deliver the required function, which leads to further losses in the customer's business.

If the capability of the company can be estimated, it is natural to think of the outcome of applying the company's capability as falling somewhere along the x-axis. The probability of achieving the performance can be represented as a normal distribution, with the expected performance represented by the mean of the normal distribution and the variation from the expected performance represented by the standard deviation.

Based on this information, the capability distribution can be superimposed onto Figure 11.1-1 as shown in Figure 11.1-2. If history shows that the company

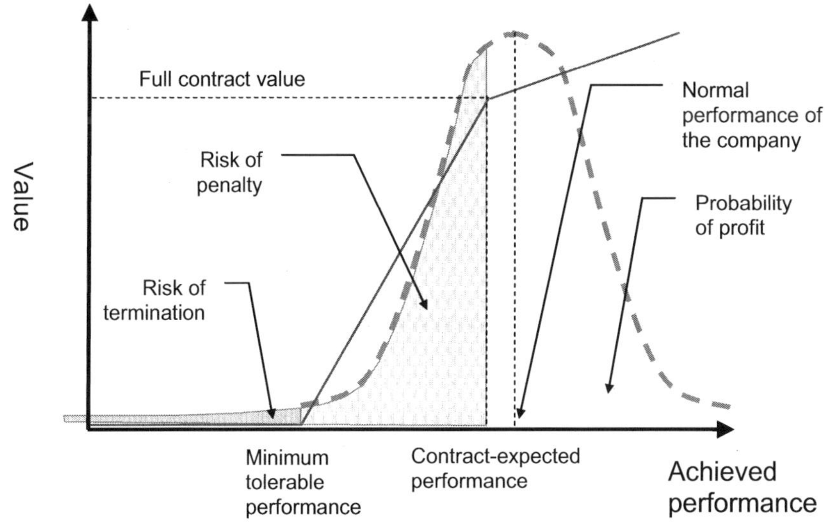

Figure 11.1-2 Company performance compared to contract requirement

normally achieves performance that is only slightly better than the contract expected performance, three probabilities in Figure 11.1-2 address such a situation. If the company achieves better than expected performance, and if the company maintains its planned expenditure, the white area under the company performance distribution is the probability of profit. If the company does not achieve its normal performance in this contract, the contractor will not be paid fully but the same planned expenditure has been spent. Therefore, the cross-shaded area represents 'risk of penalty'. Similarly, the line-shaded area represents the probability of the company not delivering even the minimum tolerable performance. The contract is terminated under this scheme. The corresponding area under the distribution curve is 'risk of termination'.

A manufacturer of a complex engineering system who enters into this kind of contract takes a lot of risks with it. The manufacturer cannot know in advance what performance can be achieve at the conceptual or design phase of the product life cycle, or even during the bidding process.

Using the 3PE model described in Section 10.3, a service enterprise can be described as a four-dimensional system model: people (X), process (Y), product (Z) and time. The new architecture covers the additional changing aspect of the service system by integrating the concepts of product, process, people and time. The four dimensions are interlinked and affect one another.

11.2 Development of performance indicators

Good management information is a significant factor in the success of a business. Consequently, an appropriate key-performance measurement system (key performance index, or KPI) is absolutely necessary to assess the system's success other than profitability and control of cash flows. KPI will be different for dissimilar organisations. For example, a university may aim its KPI at student pass or graduation rates. A retail store customer-complaints department may focus on the percentage of settled customer complaints. A transport company may measure transport efficiency across different fleets, comparing like with like. One of the KPIs for any business might be the profits generated by return customers. In any case, no matter what the specifics are, a KPI must reflect the organisation's goal.

Ideally the integrated KPI is a single-value function that reflects appropriately the influence of the individual KPIs that aggregate to form it. This idea is similar to the index of share markets such as Dow Jones or NASDAQ. Selection of constituent KPIs and their weighting will affect the accuracy of the performance assessment.

To expatiate effectiveness and efficiency of a supply chain, the following KPIs are commonly used:

- Profit.
- Lead-time performance.
- Delivery promptness.

- Waste elimination.
- Service effectiveness.
- Operational efficiency.
- Responsiveness.
- Reliability.
- Costs.
- Assets.
- Flexibility.

Before a KPI is selected, it is important to identify the organisation's goal, which in turn will affect the organisation. Development of a KPI consists of deciding business goals by using systematic methods and collecting feedback regularly to indicate whether a business is achieving its targets, whether inaccurate progress is avoided and whether customers are comfortable.

An organisation should not only control the internal management but also focus on the external environment to develop system performance as a whole. Hence an integrated KPIs development is utilised in a cohesive system with the conformity of distinct aspects of performance.

11.3 Evaluation of contract fulfilment capability

Under a performance-based contracting approach, a contractor offering systems support services must design an operations and support system that is sustainable and fit for the purpose, and demonstrates its value for money (Corsten and Gössinger, 2006). The use of an enterprise architecture to provide a starting point of the design process will reduce the risk of adopting wrong conceptual design and set the baseline ready. Using the 3PE model, the four-dimensional system model becomes: people (X), process (Y), product (Z) and time are interlinked and affecting one another. The performance of the contractor depends on the capability that the contractor has in order to deliver to the satisfaction of the customer. Mathematically, this can be expressed by the equation:

$$\{P\} = \mathrm{g}(\{\mathbf{C}\}) \tag{11.3-1}$$

where $\{\mathbf{C}\}$ is the vector describing the variations of the contractor's capabilities in the three dimensions X, Y and Z, and $[P]$ is the performance matrix discussed previously. Using this enterprise formulation, it is then possible to assess the probability of the contractor's success in meeting the contract requirements.

In each component a number of capability parameters make up the dimension as a whole. Each parameter will contribute a normal distribution of the current capability of the company. These distributions are subject to variations, due to the changing environment E:

$$\{\mathbf{X}\} = \{x_1, x_2, \ldots x_n\} \tag{11.3-2}$$

To obtain the distribution, let us examine the assessment of performance based on known capability levels. As an assessment, it is possible to estimate at three different levels:

- x_o = the most *optimistic* performance outcome due to x
- x_m = the most *likely* performance outcome due to x
- x_p = the most *pessimistic* performance outcome due to x

The probable performance outcome due to a capability can then be approximated as a normal distribution curve with:

Mean value: $$x_e = \frac{x_o + 4x_m + x_p}{6} \tag{11.3-3}$$

Variance: $$v_e = \left(\frac{x_p - x_o}{6}\right)^2 \tag{11.3-4}$$

Standard deviation: $$\sigma_e = \frac{x_p - x_o}{6} \tag{11.3-5}$$

If there are multiple capability factors, the importance of individual factor will be assigned a weight, which will integrate the effects of each factor into overall consideration.

Weighted mean $$x_e = \sum_{i=1}^{n} c_i x_i \tag{11.3-6}$$

Variance of weighted performance outcomes $$v_e = \sum_{i=1}^{n} c_i \left(\frac{x_{pi} - x_{oi}}{6}\right)^2 \tag{11.3-7}$$

The above methodology can be illustrated with an example. Please note that the values are hypothetical and could be adjusted to a more realistic value in a real case. Not all the enterprise elements in the 3PE model have been taken as reference. To put the illustration in a better context, if the targeted performance of an asset management and support service system is to aim at achieving two independent performance indicators, then

$$\{\mathbf{P}\} = \{\text{availability} = 96\%, \text{cost} = 210{,}000\}$$

Data related to the targeted system are:

- The component affected the system's availability has MTTF = T_o = 1000 days.
- The time to repair T_r varies between 1 day and 5 days with a nominal time of 2.5 days.
- Cost of each component is $10,000. The components are replaced in the annual maintenance service.

- There are 4 replacements (every 90 days) each year, so costs = \$40,000.
- Total labour cost associated with the service of this system is \$80,000 per annum.
- Overhead is 100% of labour cost.

The availability of the system due to product enterprise structure can then be computed as

$$A = \frac{T_o}{T_r + T_o} \tag{11.3-8}$$

The availabilities for the best, medium and worst cases are as shown in Table 11.3-1.

It should be noted that the availability values as shown in Table 11.3-1 are ideal availability when every other part of the system is operating in a perfect way. Suppose there are some problems with the spare parts supply chain (i.e. product/process) such that the spare parts are out of stock occasionally. Assuming that on average 20 days out of 1,000 days are lost due to waiting for spare parts, the availability after including no spare part days for the three cases is then adjusted to the values, as shown in the middle column of Table 11.3-1. Furthermore, it is assumed that inefficiency in management (i.e. people/process) such as documentation delay is estimated to cause 50 days out of 1000 days. The effect on availability is shown in the right-hand column of Table 11.3-2.

The approximated normal distribution of system availability due to a combination of organisational capability is given in Table 11.3-3. The characteristics of normal distribution are computed as

- Mean = 0.962108
- Variance = 0.0000316
- Standard deviation = 0.005625

Table 11.3-1 Availability of system due to different time to repair

Case	Time to repair (days)	Availability
Best	1	0.999001
Medium	2.5	0.997506
Worst	5	0.995025

Table 11.3-2 Availability of system after taking uncertainties into account

Case	Availability after considering out of stock	Availability after considering management inefficiency
Best	0.979021	0.949051
Medium	0.977556	0.947631
Worst	0.975124	0.945274

Table 11.3-3 Availability of system after taking uncertainties into account

Performance outcome	Availability
Optimistic	0.979021
Medium	0.962088
Pessimistic	0.945274

Table 11.3-4 Cost of system due to different overhead rates

Case	Overhead rate change in the scenario	Cost
Best	10% reduction	$192,000
Medium	0% (no change)	$200,000
Worst	15% increase	$212,000

Table 11.3-5 Total cost of service system due to staff morale

Case	Cost due to good morale (5% labour-cost reduction)	Cost due to very bad staff morale (10% labour-cost increase)
Best	$184,400	$207,200
Medium	$192,000	$216,000
Worst	$203,400	$229,200

For a desirable availability $A = 96\%$, the normalised Z variable is –0.3747 and the probability of meeting the performance is 64.6%. In other words, the probability that this improvement initiative may fail is 35.4%. This is not a brilliant outcome, but unfortunately it is an illustration of the fact that improvement projects can be quite risky.

Now let us look at the second performance criterion: cost. If the overhead rates are assumed to be different in each of the three scenarios, the cost factors are calculated as shown in Table 11.3-4.

Cost can be affected by morale of labour (i.e. people). Assume that the labour cost can be reduced by 5% if the service team has good morale. If the organisation has bad morale, the labour cost is estimated to increase by 10%. The effect of morale is listed in Table 11.3-5.

The approximated normal distribution of cost performance is computed as shown in Table 11.3-6. It should be noted that for this factor, success is defined by lowering cost values (as compared to the factor of availability, which expects an increased value).

For a desirable cost $C = \$210{,}000$, the normalised Z variable is 0.732143 and the probability of meeting the performance is 75.8%. This is a relatively good outcome which indicates that the people and its linked elements are under good control.

Table 11.3-6 Normalised performance cost-factor distribution

Performance outcome	Cost		
Optimistic	\$184,400	Mean	\$204,533
Medium	\$203,400	Variance	55,751,111
Pessimistic	\$229,200	Standard deviation	\$7,467

Last, the weighting of the two performance indicators are determined through a Delphi process or by an analytic hierarchy process. Either weighting method can provide a balanced view of the aggregated performance value. Assuming that the weighting is determined as

$$\{\mathbf{W}\} = \{\text{availability} = 0.6, \text{cost} = 0.,4\} \tag{11.3-9}$$

the combined performance distribution has an expected mean of 0.69482. The standard deviation of the normalised distribution is 1.0, so the risk of the service improvement project not meeting the expected target performance is assessed as 24.4% (i.e. the probability of success is 75.6%).

It is up to the decision maker of the company to decide whether the risk is too high (non-acceptance) or reasonable (worth taking). If this risk is not acceptable, further investigation can be carried out to explore what the company can improve apart from the above; for example, by improving the process (e.g. process mapping) or spare parts stock control (e.g. eliminating stock out).

11.4 Failure analysis

Failure analysis can take many forms. FMCA is the most well-known system for new products and engineering changes. The computation of failure probability is the basic requirement for understanding the risk in different failure patterns. Failure probabilities can be estimated by exhausting all possible failure scenarios. We use a simple brake-system example to illustrate how such probabilities can be assessed.

In a normal situation, the brake shoes are relaxed and the tyres are free to rotate; thus the car has no braking effect. Braking is actioned by driver's foot pressing onto the pedal. Depending on the extent of pressing, oil in the two master cylinders is squeezed out to the brake lines. The brake lines are connected to the brake shoes and are operating at extremely high pressure when the hydraulic system is activated. The oil passes through the lines and presses onto the brake shoes. The brake shoes have a lever mechanism that expands to push onto the tyre-wheels. The rotation of the tyres is therefore counteracted, which causes unmatching surface speed between the tyres and the road. Friction between the tyres and the road then slows down the car (Figure 11.4-1).

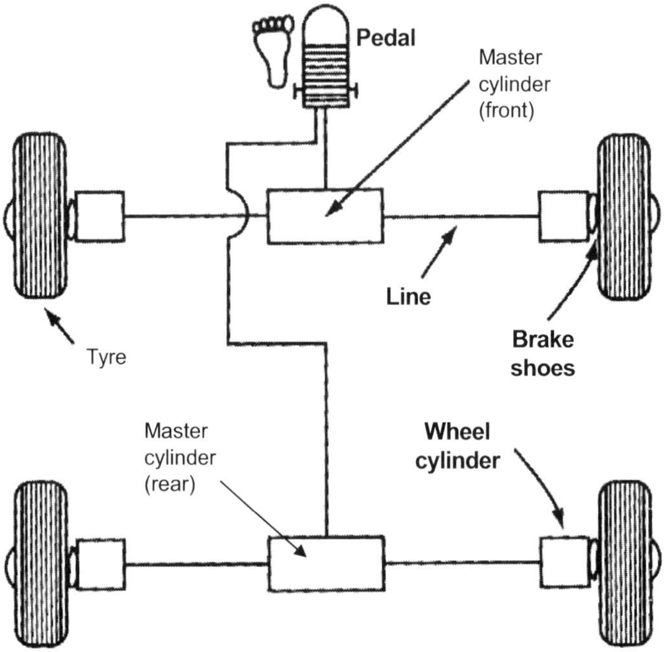

Figure 11.4-1 Schematic of dual hydraulic braking system

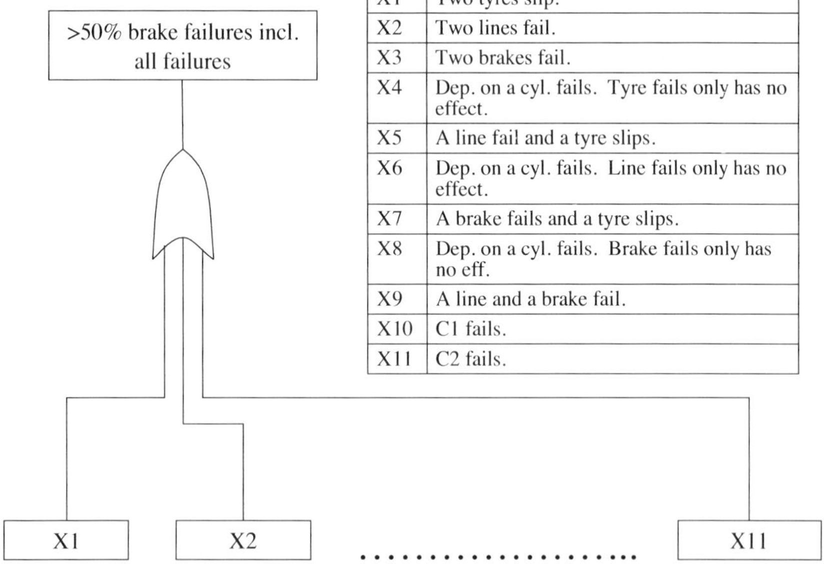

X1	Two tyres slip.
X2	Two lines fail.
X3	Two brakes fail.
X4	Dep. on a cyl. fails. Tyre fails only has no effect.
X5	A line fail and a tyre slips.
X6	Dep. on a cyl. fails. Line fails only has no effect.
X7	A brake fails and a tyre slips.
X8	Dep. on a cyl. fails. Brake fails only has no eff.
X9	A line and a brake fail.
X10	C1 fails.
X11	C2 fails.

Figure 11.4-2 Failure modes

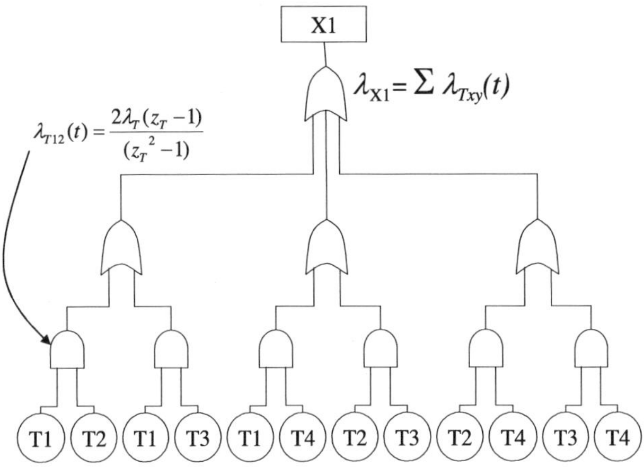

Figure 11.4-3 Failure rates of failure mode X1 (any two-tyre slip)

The following component failure rates are given:

Master cylinder = 0.002
Line = 0.00005
Brake shoe = 0.005
Tire friction = 0.01

If the system will fail when it loses 50% of its functionality, what is the probability of failure? To answer this question, first of all we need to define the failures. Figure 11.4-2 shows all possible failure modes. For each of the failure modes, an individual fault tree can be developed and the failure rates can be computed with the given information (Figure 11.4-3).

Failure mode X5 is more complicit, as shown in Figure 11.4-4. The failure rates of the failure modes in this example (i.e. cases causing >50% brake failures) are shown

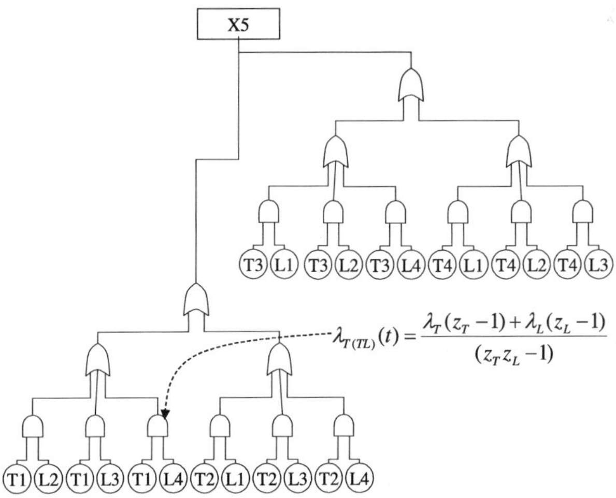

Figure 11.4-4 Failure mode X5 (a line fails and tyre in another section slips)

Table 11.4-1 **Failure rates of all failure modes**

	T1	T2	T3	T4	C1	C2	L1	L2	L3	L4	B1	B2	B3	B4
T1	Void	0.007746	0.007746	0.007746				4.99539E-05	4.99539E-05	4.99539E-05		0.00447835	0.004478	0.004478
T2		Void	0.007746	0.007746			4.99539E-05		4.99539E-05	4.99539E-05	0.004478		0.004478	0.004478
T3			Void	0.007746			4.99539E-05	4.99539E-05		4.99539E-05	0.004478	0.00447835		0.004478
T4				Void			4.99539E-05	4.99539E-05	4.99539E-05		0.004478	0.00447835	0.004478	
C1					0.018548									
C2						0.018548								
L1							Void	4.96277E-07	4.96277E-07	4.96277E-07		3.4769E-05	3.48E-05	3.48E-05
L2								Void	4.96277E-07	4.96277E-07	3.48E-05		3.48E-05	3.48E-05
L3									Void	4.96277E-07	3.48E-05	3.4769E-05		3.48E-05
L4										Void	3.48E-05	3.4769E-05	3.48E-05	
B1											Void	0.00282367	0.002824	0.002824
B2												Void	0.002824	0.002824
B3													Void	0.002824
B4														Void

Table 11.4-2 Computed probability of brake failure

X1	Two tyres slip	0.046476
X2	Two lines fail	2.98E-06
X3	Two brakes fail	0.016942
X4	Dep. on a cyl. fails. Tyre fails but has no effect.	0
X5	A line fails and a tyre slips	0.000599
X6	Dep. on a cyl. fails. Line fails but has no effect.	0
X7	A brake fails and a tyre slips.	0.05374
X8	Dep. on a cyl. fails. Brake fails but has no effect.	0
X9	A line and a brake fail	0.000104
X10	C1 fails	0.018548
X11	C2 fails	0.018548
Total (<50% brake)		0.154962

and

	Tyre	Cylinder	Line	Brake
l	0.01	0.02	0.00005	0.005
z	1.581976707	1.156518	200.5004	2.541494
l (event)	0.004422938	0.018548	4.96E-07	0.002824
		Total (no brake)		0.025796

in Table 11.4-1. Combining the failures, and assuming time t is 100 hours, the total failure probability is given in Table 11.4-2.

11.5 Contractual risks of in-service assets

Services to in-service assets are often constrained by the availability of the asset during operations. There are also upgrade requirements that make it difficult to estimate the risks accurately (Kutsch and Hall, 2005). We use a shipping company example to illustrate the complexity of such service contracts.

A ship-building company SB is selling ocean liners to large shipping companies. In recent years, their business has changed such that they are not only selling ships but also supporting the operations and upgrades of the ocean liners they sell.

One of the shipping companies LC requests for tender on a service contract that covers the maintenance and upgrade services of five (5) ships that the SB just sold. The contract will last for 30 years. This is good long term business for the builder. The ships will carry containers among five ports of Australia and three cities in East Asia: Singapore, Hong Kong and Yokohama.

Table 11.5-1 Probabilities of ship having service in ports

Ship ID	A	B	C	D	E
Sydney	0.1	0.3	0.1	0.1	0.1
Brisbane	0.1	0.2	0.3	0.1	0.1
Melbourne	0.3	0.1	0.1	0.1	0.2
Adelaide	0.2	0.1	0.1	0.3	0.1
Perth	0.1	0.1	0.1	0.1	0.3
Singapore	0.1	0.0	0.1	0.1	0.1
Hong Kong	0.1	0.1	0.1	0.1	0.0
Yokohama	0.0	0.1	0.1	0.1	0.1
TOTAL	1.0	1.0	1.0	1.0	1.0

Table 11.5-2 Probabilities of port facilities setup cost

Setup cost in $M	2	4	6	8	10	TTL
Sydney	0.1	0.3	0.3	0.2	0.1	1.0
Brisbane	0.1	0.3	0.4	0.1	0.1	1.0
Melbourne	0.2	0.2	0.3	0.2	0.1	1.0
Adelaide	0.0	0.1	0.4	0.4	0.1	1.0
Perth	0.1	0.2	0.3	0.2	0.2	1.0
Singapore	0.0	0.1	0.1	0.3	0.5	1.0
Hong Kong	0.1	0.1	0.2	0.2	0.4	1.0
Yokohama	0.0	0.0	0.1	0.3	0.6	1.0

The contract conditions are discussed in the following paragraphs.

SB is responsible for maintenance and services to the ships whenever there is a need, including emergency services required at the three overseas cities. The probabilities of a ship having maintenance services at the ports are shown in Table 11.5-1. The cost of setting up support is a one-off outlay. The probabilities of this outlay are shown in Table 11.5-2. An ongoing cost of 10% setup cost is required if the facility continues to be used for this support.

The profit of operating a ship is $0.1M per day. Hence, SB does not want to let the ship stop for maintenance. However, SB guarantees that any maintenance, system upgrade or emergency work will not exceed a maximum of 10 days. If the service takes more than 10 days, SB will compensate LC for the loss by $0.03M per day. The duration of maintenance service depends on the capability of the ports (Table 11.5-3). The cost of maintenance services also varies (Table 11.5-4).

SB has the option of using subcontractors at the overseas ports. The cost of subcontracting is 10% higher than those shown in Table 11.5-4 but SB can save the setup cost (e.g. $2.2k per day in Hong Kong) for a probability of 0.1. SB can subcontract or withdraw at any time according to an annual review. However, if SB initially sets up the facility and later subcontracts, the setup cost will not be returned. SB will need

Table 11.5-3 Probabilities of maintenance time

Time in days	2	4	6	8	10	12	14	TTL
Sydney	0.0	0.2	0.3	0.3	0.05	0.05	0.1	1.0
Brisbane	0.1	0.2	0.3	0.3	0.05	0.05	0.0	1.0
Melbourne	0.1	0.2	0.4	0.2	0.1	0.0	0.0	1.0
Adelaide	0.05	0.1	0.4	0.3	0.1	0.05	0.0	1.0
Perth	0.1	0.1	0.3	0.3	0.2	0.0	0.0	1.0
Singapore	0.0	0.2	0.3	0.3	0.1	0.1	0.0	1.0
Hong Kong	0.1	0.2	0.3	0.3	0.1	0.0	0.0	1.0
Yokohama	0.0	0.4	0.2	0.1	0.1	0.1	0.1	1.0

Table 11.5-4 Probabilities of maintenance cost

Cost $k per day	2	4	6	8	10	12	14	TTL
Sydney	0.05	0.2	0.3	0.3	0.1	0.05	0.0	1.0
Brisbane	0.1	0.2	0.3	0.3	0.05	0.05	0.0	1.0
Melbourne	0.1	0.2	0.4	0.2	0.1	0.0	0.0	1.0
Adelaide	0.05	0.1	0.4	0.3	0.1	0.05	0.0	1.0
Perth	0.1	0.1	0.3	0.3	0.2	0.0	0.0	1.0
Singapore	0.0	0.4	0.3	0.1	0.1	0.1	0.0	1.0
Hong Kong	0.1	0.4	0.3	0.1	0.1	0.0	0.0	1.0
Yokohama	0.0	0.2	0.2	0.3	0.1	0.1	0.1	1.0

to re-setup if SB changes its mind (i.e. SB will incur the same setup cost again if SB cannot provide the services overseas for 1 year).

The ships are controlled by modern IT systems which change frequently. Hence SB will need to upgrade these systems as required. The time of upgrade depends on many factors and can be expressed as probabilities (Table 11.5-5). Over the 30 years of contract, SB expects to upgrade several times. The probability of upgrade is assessed from the last upgrade (or from new installation).

When an upgrade occurs, the cost depends on the technology requirements and state of the asset. The probabilities of upgrade cost in $M of current value can be

Table 11.5-5 Probabilities of upgrade timing

Year	Probability of upgrade occurring in the year
1	0.1
2	0.1
3	0.3
4	0.4
5	0.1

Table 11.5-6 Probabilities of upgrade cost

Cost (in $M)	Probability of upgrade cost
1.0	0.1
2.0	0.2
3.0	0.2
4.0	0.3
5.0	0.2

expressed as a probability function (Table 11.5-6). The annual marginal rate of return is 10%.

We now ask, what is the bidding price of the contract (in net present value) with only 5% chance of running a loss?

There are many ways to solve this problem. The method presented here is based on analytical statistics. Alternative methods such as simulation can also be used if the simulation spreadsheet is set up properly. The analytical statistics method assumes that the tabulated statistics can be approximated to a normal distribution.

Recall the following statistical definitions:

Definition 1. If X is a discrete random variable with probability mass function $p(x)$ then the expected value is given by:

$$E(X) = \sum_i x_i p(x_i) \qquad (11.5\text{-}1)$$

Definition 2. If X is a discrete random variable with probability mass function $p(x)$ then the variance is given by:

$$Var(X) = \sum_i (x_i)^2 p(x_i) - (E(X))^2 \qquad (11.5\text{-}2)$$

The standard deviation $Std(X)$ can then be taken as the square root of $Var(X)$.

Using (1) and tables 2 and 4, and noting the cost of using subcontractors in overseas cities as per the figures with brackets, the expected and standard deviation values of the set up and daily maintenance costs respectively per city are:

First we note that using (1) and Table 5, the expected length and standard deviation of time for an upgrade are 3.3 and 1.1 years respectively. Similarly, from Table 6, the expected cost and standard deviation of this upgrade are \$3.3M and \$1.62M respectively. This can be written in the following table.

We also calculate the expected amount that the company will pay in a given year from maintenance time longer than 10 days. In order to do this, we use the probabilities from tables 1 and 2. So, for instance, the mean amount that ship A will be compensated for stopping in Sydney will be:

$$0.1 \times 0.03(0.05 \times 2 + 0.1 \times 4) = \$1.5\text{k} \qquad (11.5\text{-}3)$$

If we make this calculation for all cities and then add all the totals, we declare that the mean amount of compensation that ship A will be paid from maintenance times longer than 10 days is $3k. Using this same approach, we calculate the amounts for all the ships in Table 11.5-3. We can also determine the standard deviation of the compensation amount per ship (Table 11.5-4).

Now we can explain the reasoning for tabulating all of the standard deviations. Suppose that we want to estimate an upper bound with probability 84.1% for the set-up cost. To do this, we simply take the sum of all of the expected values in Table 1 and add 1 multiple of the standard deviation from each city, which is equivalent to the total amount for standard deviation. We also recall that a factor of 10% of set-up cost is also included as an ongoing cost.

Last, because the duration of the contract bid 30 years and the marginal rate of return is 10%, $A(P,i,n) = 0.12414$.

In short, the reasoning is based on simply looking at the expected values and ignoring the standard deviations.

We also note that we have the option of using subcontractors in the overseas cities. This saves on the set-up cost but means an extra 10% on the daily maintenance cost of those cities. We also note that the expected upgrade costs for the duration of the contract comes to $30M (from 3.3/3.3 × 30).

Now, without subcontractors, the total annual setup and daily maintenance cost is given by

$$((54.2+k18.76)\times 1.1+30)\times 0.12414+18.8+k8 \qquad (11.5\text{-}4)$$

whereas with contractors, the annual setup and daily maintenance cost is given by

$$((29.4+k12.77)\times 1.1+30)\times 0.12414+19.5+k8.3 \qquad (11.5\text{-}5)$$

If we put $k = 0$, then we note that (3) and (4) give amounts of $29.9M and $27.2M respectively. Thus in this instance we will choose to use subcontractors for the overseas cities.

Furthermore, from Table 3, the total expected amount that will be paid to all ships for maintenance time longer than 10 days is $0.0249M.

Thus the total expected annual cost is calculated as: 27.6 + 0.0249 ≈ $27.22M. If we then decide that we want an annual profit of 10% of the cost, we make a bid of $29.94M per year. Hence if we make a bid of $29.94M we could expect a profit of 10% on our invested costs.

However, we would also like to assess whether this is a risky deal for the company. A way to determine this is by analysing the distribution of the total cost. We note that our two main costs will be the setup cost (Table 2) and the upgrade costs as determined by the frequency (Tables 5 and 6) of these upgrades. Using the probabilities from these tables, we run a simulation 100 times to obtain the setup costs and upgrade costs for the cases with and without subcontractors.

For the case with contractors, we find that the total amount from the setup and upgrade costs for the duration of the 30-year contract would be $67.23M with a standard deviation of $6.16M. Without subcontractors, the entire amount for the 30-year contract would be \$87.3M and with a standard deviation of \$5.08M. For the benefit of the reader, we also present the respective histograms of these scenarios.

Next, we must determine whether it is better go with or without subcontractors. Similar to the above, the total annual setup and daily maintenance cost is given by

$$(87.3+k18.76)\times 0.12414+18.8+k8 \tag{11.5-6}$$

where with contractors, the annual setup and daily maintenance cost is given by:

$$(67.23+k12.77)\times 0.12414+19.5+k8.3 \tag{11.5-7}$$

If we now let $k = 1.65$, we would then obtain an upper bound, with a 95% probability, for those costs. Without subcontractors, this comes to \$48.15M and with subcontractors, this comes to \$45.25M. Thus, the company will use subcontractors. If we then also consider the expected value of \$0.0249M of compensation that will be paid to the shipping company for maintenance times taking longer than 10 days, and factoring in a 10% profit then we should make a bid of \$49.8M per day.

Some other observations can be made:

- The simulation results did not greatly disagree with the expected value costs. Without contractors, the difference was \$89.62M – \$87.3M = \$2.32M; with subcontractors, the difference was \$67.23M – \$62.34M = \$4.89M.
- In all cases, unless otherwise indicated, the expected maintenance costs and time were on an annual basis. We have also assumed that the given costs, time periods etc. are normally distributed.
- The contract will be flexible enough so that if the shipping company decides to operate with more ships, these can be accommodated on a ship-per-ship basis without incurring undue penalties.

11.6 In-service risks

Many events happen during in-service periods. The effect of these events on the service system can be visualised by an event tree (Ericson, 2005), which is:

- A graphical representation of the logic model that identifies and quantifies the possible outcomes following an initiating event.
- It provides an inductive, step-by-step assessment leading to multiple possible outcomes.

In general, an event tree can be constructed as shown in Figure 11.6-1. We use an example to illustrate how an event tree can be constructed. The process to be modelled is the delivery of goods to warehouse (Figure 11.6-2). Delay can be due to:

- Traffic congestion.
- Vehicle breakdown.
- Picking or packing error.

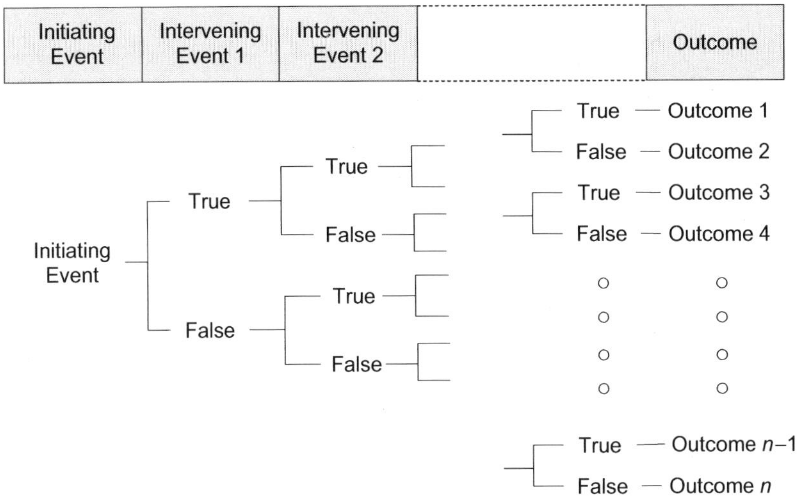

Figure 11.6-1 General even-tree construction

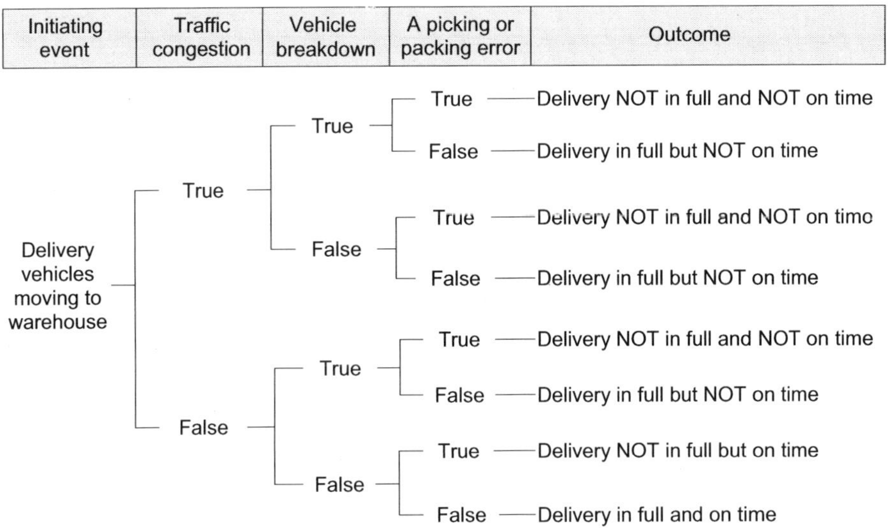

Figure 11.6-2 Event tree for goods delivery to warehouse

Next, we can insert probabilities of the events and compute that probability of fail to deliver in full and not in time is 0.00703%; fail to deliver in time but delivered in full is 70.29%; and fail to deliver in full but delivered in time is 0.00297% (Figure 11.6-3). The probability of success is 29.70% and probability of delivery not in full is 0.146 (from Table 11.6-1 to Table 11.6-4).

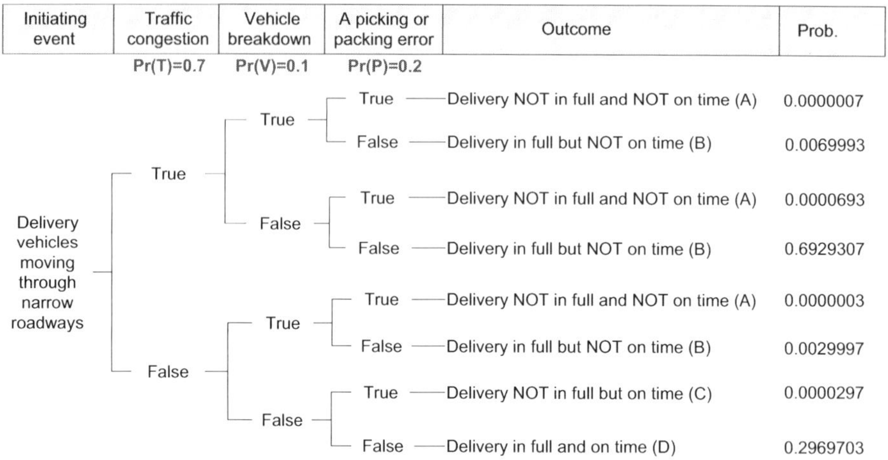

Figure 11.6-3 Event tree for goods delivery to warehouse

Table 11.6-1 Expected and standard deviation of set up and maintenance costs

	Setup ($M)		Daily maintenance ($K)	
City	**E**	**S.D.**	**E**	**S.D.**
Sydney	5.8	2.27	6.7	2.39
Brisbane	5.6	2.15	6.3	2.47
Melbourne	5.6	2.5	6	2.19
Adelaide	7	1.61	6.3	3.59
Perth	5.4	4.24	6.8	2.4
Singapore	8.4	1.96	6.4 (7.04)	2.65 (2.92)
Hong Kong	7.4	2.69	5.4 (5.94)	2.2 (2.42)
Yokahama	9	1.34	7.6 (8.36)	3.98 (4.38)
Total	54.2	18.76	51.5 (53.44)	21.87 (22.76)
Yearly Total	54.2	18.76	18.8M (19.5M)	8M (8.3M)

Table 11.6-2 Frequency and cost-upgrade values

Time		**Cost**	
(years)		**($M)**	
E	S.D	E	S.D
3.3	1.1	3.3	1.61

Table 11.6-3 Expected compensation amount per ship ($k)

City	Ship A	B	C	D	E
Sydney	1.5	4.5	1.5	1.5	1.5
Brisbane	0.3	0.6	0.9	0.3	0.3
Melbourne	0	0	0	0	0
Adelaide	0.6	0.3	0.3	0.9	0.3
Perth	0	0	0	0	0
Singapore	0.6	0	0.6	0.6	0.6
Hong Kong	0	0	0	0	0
Yokahama	0	1.8	1.8	1.8	1.8
Total	3	7.2	5.1	5.1	4.5

Table 11.6-4 Expected compensation amount per ship ($k)

City	Ship A	B	C	D	E
Sydney	10.3	14.3	10.8	10.8	10.8
Brisbane	3	4.2	5.2	3	3
Melbourne	0	0	0	0	0
Adelaide	4.2	3	3	5.2	3
Perth	0	0	0	0	0
Singapore	4.2	0	4.2	4.2	4.2
Hong Kong	0	0	0	0	0
Yokahama	0	11.2	11.2	11.2	11.2
Total	19.2	29.7	31.4	34.4	33.2

References

Corsten, H., Gössinger, R., 2006. Output flexibility of service enterprises — an analysis based on production theory. Int. J. Prod. Econ. 104 (2), 296–307.

Defence Materiel Organisation, 2007. Performance Based Contracting Handbook – Guiding Principles and Performance Framework, Version 2.0, Department of Defence, Australia, February, pp. 138. http://www.defence.gov.au/dmo/asd/publications/asd_pbc_v2.pdf.

Ericson, C.A., 2005. Event Tree Analysis, in Hazard Analysis Techniques for System Safety. John Wiley & Sons, Chapter 12, ISBN: 9780471720195.

Kutsch, E., Hall, M., 2005. Intervening conditions on the management of project risk: Dealing with uncentainty in information technology projects. Int. J. Proj. Manage. 23, 591–599.

Continuous improvement

12

Chapter Outline

12.1 The need for continuous improvement

The volatility of current business environment requires companies to adapt to new processes and systems to satisfy customer requirements as well as remain competitive. Business enterprise is inherently a complex entity. There are many risks involved in the changes that an enterprise must go through in order to transform itself into a more competitive form (Rouse, 2005). Typical risks include collaboration, confidentiality, intellectual property, transfer of goods, conflicts, opportunity loss, product liability and others. Inappropriate actions can be taken if the information is out of date or the employee performance can be seriously affected. Issues of enterprise change should be dealt with in a systematic fashion, supported by a methodology that assists the whole of the enterprise to transform.

Enterprise modelling is best used to analyse the business in both manufacturing and service sectors in terms of complexities and context. Enterprise-reference architectures provide structures by which to understand enterprise activities such as promoting planning, reducing risk, implementing new standard operating procedures and controlling and rationalising manufacturing facilities. These types of enterprise-engineering research share a common starting point in that stepwise, multidimensional enterprise-modelling and -design methodologies have been applied. The rationale behind the use of enterprise-engineering methodologies to guide these developments is the minimisation of the impact of uncertainty upon enterprise performances as well as other associated processes.

However, the present enterprise-modelling methodologies contain limitations when they are applied to the modelling of enterprise transformation. One commonly used methodology to model change is six-sigma methodology. The process focus of six sigma can benefit an enterprise by providing the means to develop a road map and to initiate the enterprise transformation that is required and could become more culturally ingrained within the organisation.

12.2 Six-sigma processes

Everything in a business can be viewed as a process. A manufacturing enterprise receives an order, schedules the production, builds the product, delivers the product and receives payment. A service enterprise receives a customer request, schedules a customer appointment, delivers the service and receives payment. For process improvement, six sigma is a powerful tool for achieving and sustaining operational and service excellence. The original focus of six sigma was on manufacturing, but today it is widely accepted in both service and transactional processes (Smith, 2001).

Six sigma is a methodology for process improvement through reduced variability and the elimination of defects. Six sigma addresses system deficiencies by using data to make decisions and formulate data-driven solutions. The tool set, which is readily available, is a collection of well-known methods and techniques. Intuition may be used to brainstorm, but all decisions are made using measurable data benchmarked against a set of critical to satisfaction (CTS) criteria, hence achieving measurable financial returns to the bottom line of an organisation.

Six sigma is structured as a sequence of processes. DMAIC and DMADV are two streams of processes that deal with specific changes in organisations. DMAIC methodology is used to improve existing products or processes that are not performing to target or not meeting customer expectations. DMADV is used when the process or product does not already exist but is required or when an existing process can not be optimised. Both streams consist of five steps (Table 12.2-1). Six-sigma processes are best represented as a cyclic-spiral continuous-improvement methodology (Figure 12.2-1).

Six sigma is a tool that can create a new process or improve existing processes. Its main scope of application has involved making incremental changes to particular

Table 12.2-1 DMAIC and DMADV processes

	DMAIC		DMADV
D	Define the project goals and customer (internal and external) CTS deliverables.	D	Define the project goals and customer (internal and external) CTS deliverables.
M	Measure the process CTS deliverables to determine current performance using verified measurements.	M	Measure and determine customer needs and specifications.
A	Analyse and determine the root cause(s) of the defects.	A	Analyse and characterise the process options to meet the customer needs.
I	Improve the process by eliminating defects.	D	Design (detailed) and optimise the process to meet the customer needs.
C	Control future process performance.	V	Verify the design performance and ability to meet customer needs.

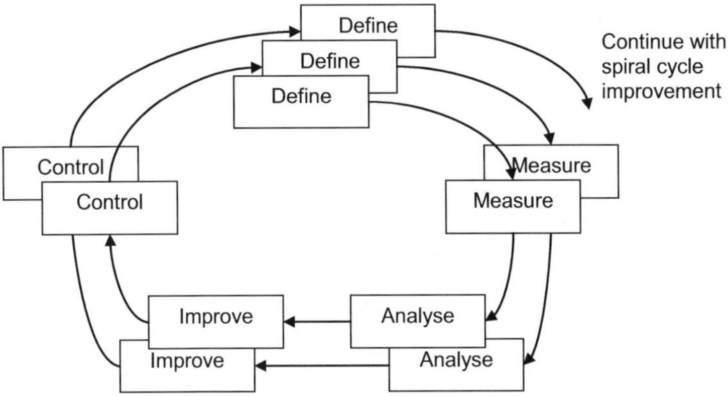

Figure 12.2-1 Spiral six-sigma improvement cycle

segments of an enterprise such as cultural change, customer focus, process elements and statistical methods of improvement.

12.3 Enterprise transformation modelling

Six sigma needs a clear, transparent, integrated definition and description of the processes of the enterprise for it to optimise and operate successfully. Because enterprise architecture design approach uses tools and methods that take a bird's-eye view of an enterprise, a methodology that combines enterprise-modelling and six-sigma approaches can maximise global-improvement outcomes during enterprise-model transformation.

Enterprise modelling provides a total-enterprise view of processes, resources and technology. The levels of integration, duplication or inefficiency in various processes may be derived through development of the 'as-is' enterprise model. Figure 12.3-1 shows a generic roadmap for enterprise-model transformation. The existing business is represented with enterprise model M_1 (i.e. an 'as-is' model). The forecast requirement is for an enterprise model as represented by enterprise model M_n (i.e. a 'to-be' model).

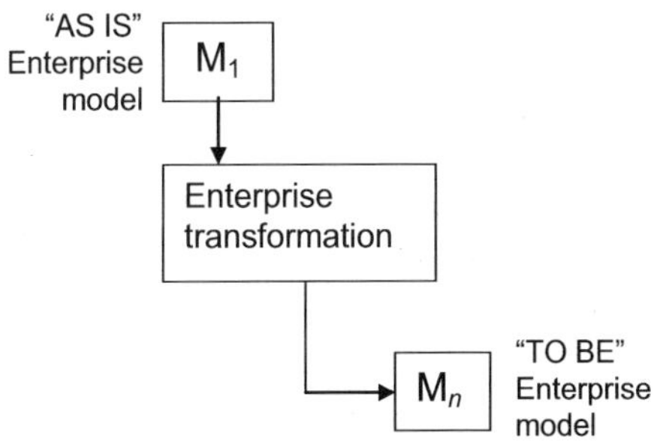

Figure 12.3-1 Enterprise-model transformations

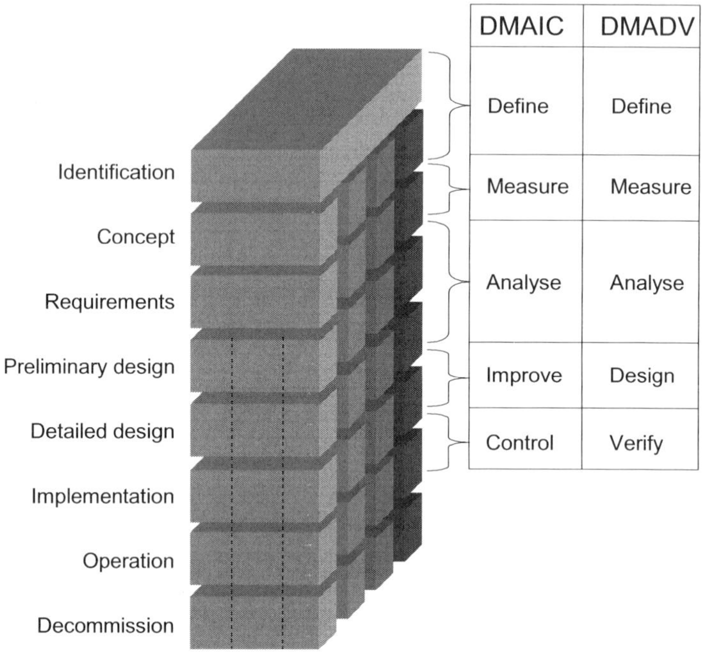

Figure 12.3-2 GERAM alignment with six sigma

Enterprise-modelling-based transformation can be compared to six-sigma methodology with the life-cycle views of GERAM (IFIP–IFAC Task Force, 1999). Figure 12.3-2 shows how each of the five steps in six sigma would align with the phases of the enterprise-life cycle view.

In an as-is enterprise model, the entire enterprise development is described as a snapshot of processes that consists of overlaying the four views. Irrespective of which phase of a process has been developed, each process is explored in detail as if it has been fully designed, implemented, operated and decommissioned. The enterprise is then regarded as fully described. If there is no change in identification and concept, which are often influenced by external parameters, the enterprise will progress through its life cycle. Actual performance for each process may be benchmarked against targets, thus highlighting processes with short fall performance. If the enterprise chooses, these processes can then be identified as candidates for six-sigma projects in which the processes would be improved by focusing either on variability using the DMAIC process or redesigned using the DMADV process.

Six sigma has a narrow focus that confines objectives in specific areas that maximise project support, with the ultimate aim of successful conclusion. Transformation projects are selected by identifying what is critical to satisfaction (CTS) for the customer and what areas of the business are deficient. Six-sigma methodology contains tools to identify the measurable deliverables that are required for customer satisfaction; it

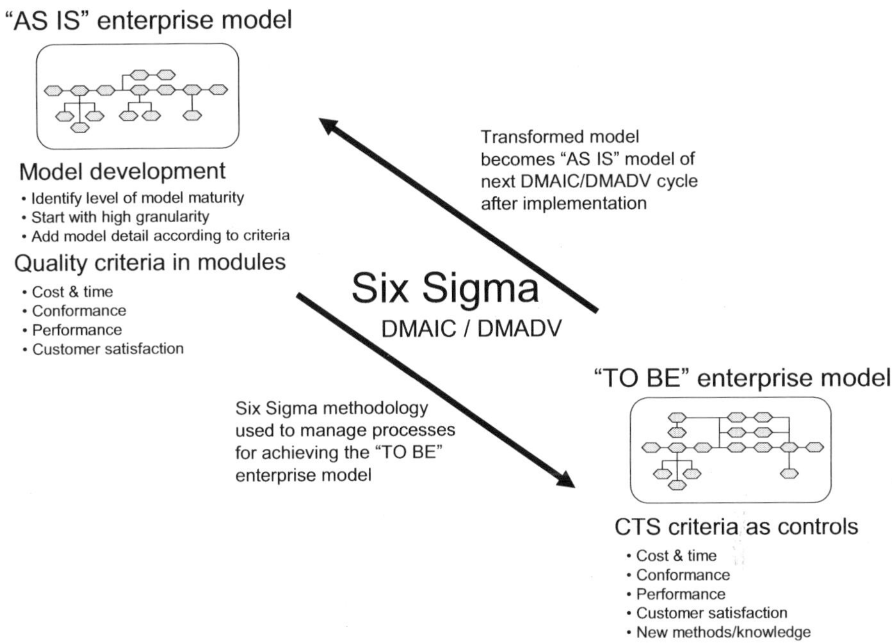

Figure 12.3-3 Transition plan with six sigma

also identifies areas, processes or inputs that influence these deliverables. Identifying and prioritising these areas, processes or inputs against relevant criteria is one method that can be used to select six-sigma projects that will lead to effective and progressive enterprise transformation.

The as-is enterprise model can have varying levels of model maturity and granularity. Using six sigma, the business process engineer can focus on the relevant areas and develop the necessary details within the project scope. Thus, as shown in Figure 12.3-3, a six-sigma project will take the as-is enterprise model to represent the current real-world condition and then establish the relevant CTS criteria. This intuitive alignment between the two approaches encourages the integration of six-sigma methodology into the generation of new enterprise models. As the enterprise model is measured and analysed, six-sigma methodology continues to refine CTS aspects that will ultimately change the concept and requirements of the enterprise by shifting the context of who the customer is. These changes in context or view will enable the creation or improvement processes that will, in turn, result in the to-be enterprise model.

The added benefit of using six sigma is that a set of controls for the process and the enterprise can be established and should be included in the enterprise model. With dynamic market demands, the characteristics that are today considered critical to quality may not necessarily be meaningful tomorrow. All CTSs should be critically examined at all times and refined as necessary. For this reason, the control plan developed from the final stage of the six-sigma project should provide measures that indicate the performance and continued relevance of the processes within the transformed enterprise model (Rouse, 2005). These measures can be effective in maintaining the enterprise model and the relevance of the business to changing conditions.

The issue of enterprise transformation is the uncertainty of future enterprise models. In other words, M_n may not be fully described. This means that the roadmap for transformation from M_1 to M_n may not be defined until very late in the project, which would cause significant disruption of the transformation process and turbulence in the enterprise. Change from M_1 to M_n is normally significant because an organisation will normally seek advice from a wide variety of resources to achieve the new design. This change is too abrupt, so it will impose significant organisational turmoil and reduce the ability of the enterprise to compete against other organisations.

Six-sigma process has the advantage of self-adjusting the system to suit the need for changes. The question is how to integrate six-sigma tools as the method to make changes that are more than just continuous optimisations of a process. Therefore, it is necessary to link six sigma with enterprise modelling, in order to transform current-state as-is enterprise models to future-state to-be enterprise models via a clear road map.

If we follow the six-sigma methodology, we will focus the transformation on process elements that have measurable CTS indicators that are relevant to internal business and external customer requirements. Therefore, as discussed in Figure 6, the CTSs form the foundation for decision making when redesigning or improving the process. The same indicators may serve as controls to ensure continued performance of the new process or as indicators of when the new process is no longer relevant. Six-sigma projects close the transformation loop by mapping CTS-driven design requirements of the revised enterprise model.

Six sigma produces data-driven solutions with relevant new and/or improved processes for the next to-be enterprise model. It is therefore imperative that the initial as-is enterprise model contains accurate and relevant CTS information. It may be necessary to redefine the as-is enterprise model to ensure that the CTS indicators and measurables are included and that the enterprise model does in fact represent the real life situation. This is in essence the first step (define) of the six-sigma project. The decision for any redesign efforts over traditional continuous improvement depends on a number of variables including risk, available technology and resources, cost, customer demand, time and complexity etc. Discerning these variables is part of the six-sigma methodology and is covered in the define stage in both DMAIC- and DMADV-based projects.

Instead of including an abrupt change between two major enterprise models, the new transition methodology divides the gap into many manageable steps (Figure 12.3-4). The intermediate to-be enterprise models M_1 ($1 < i < n$) in these steps are developed through six-sigma projects. The changes are simple enough to be redefined. However, we emphasise the need to make the changes as smooth and manageable as possible to minimise impact. The future model M_n is defined to ensure a consistent approach or direction that governs each intermediate state to be processed by six-sigma projects or combinations of many grouped projects.

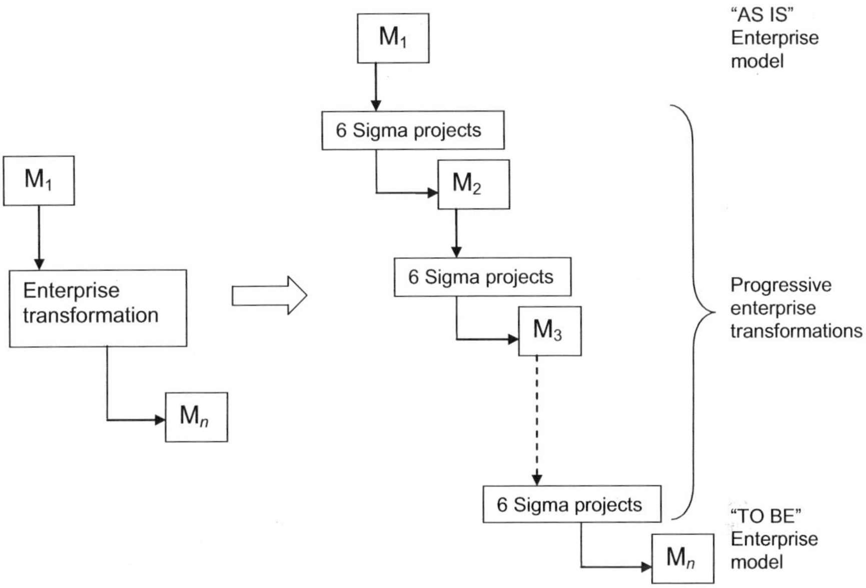

Figure 12.3-4 Enterprise model transformations with progressive six-sigma-driven changes

The road map developed for enterprise-model transformation can be context-related. The detail of each progressive enterprise model on the roadmap of enterprise transformation to M_n can be derived from six-sigma projects with emphasis on the appropriate M_1 design goals towards the ultimate state. This six-sigma-driven transformation thus becomes an iterative and progressive process of making changes that lead to the larger enterprise-model transformation in a manner that is similar to climbing a mountain. As the summit is ascended one stage at a time, successive new intermediate enterprise models may be developed until the summit can be reached in a final project. This progressive method of transformation is a logical approach for enterprise-model development that could become culturally more ingrained within an organisation that is already embracing six sigma for need-based, ongoing enterprise-model transformation (EMT).

From a more generic perspective, combining enterprise modelling with six sigma provides many benefits. Six-sigma projects benefit from a big picture of a process life cycle that includes respective views or contexts. Enterprise modelling gains a tool set with methodology for transforming an as-is enterprise model to the to-be state. Within any stage of the life cycle, the proposed alignment of the six-sigma methodology to the generic enterprise-architecture views remains the same. The changes from M_1 to $M_1 + 1$ is, by methodological design, incremental and manageable. The same tools and methods remain applicable during each of the steps used in transforming the different views of the enterprise model. The enterprise does not expect surprising disruptions to this process and hence can progress smoothly through the enterprise-model transformation.

References

IFIP–IFAC Task Force, 1999. GERAM: Generalised Enterprise Reference Architecture and Methodology. Version 1.6.3, Annex to ISO WD15704, Requirements for enterprise-reference architectures and methodologies, Mar 1, p. 39.

Rouse, W.B., 2005. A Theory of Enterprise Transformation Systems Engineering. Wiley InterScience 8 (4), 279–295.

Smith, L.R., 2001. Six sigma and the evolution of quality in product development. Six Sigma Forum Magazine 1 (1), 28–35.

Final remarks

13

Chapter Outline

13.1 Intangible assets in support system lifecycle

We recall in Chapter 3 that systems engineering principles should be applied to the full life cycle of complex engineering systems development. The essential part of this development is the accumulation of intangible asset to form the basis for integrating the activities in the early stages to later stages of the system. Figure 13.1-1 shows a representation of the process.

It is generally accepted that the ratio of cost of acquisition to cost of support and sustenance is about 1–4. Therefore, it is important not to lose sight of the need for service engineering. Adoption of the systems engineering approach can give the stakeholders a clear picture of what the 'real-life' industry project is and how new system design is developed meeting user expectation and requirements. Many influential factors exist in 'real life', for example climate change, oil price uncertainties and population growth.

However, no two systems acquisition and support projects are the same because each system needs to meet different user requirements. Due to the highly individualised nature of the acquisition and support of complex engineering systems, it is an extremely knowledge-intensive, labour-rich business that requires the development of new systems in the form of design certification, digital manufacturing, integrated testing, customer support centres, information infrastructure, human organisations, multi-disciplinary engineering teams, supply chains and distribution networks. Operations of these engineering units in a globally distributed environment present significant challenges to the engineering systems developers.

The complexity of these engineering products means the need for a comprehensive set of knowledge that covers the complete systems engineering life cycle from design to operations support, mid-life upgrades, services and logistics, maintenance and repairs.

To address these issues, business models are outcome based that drives contractors to take up more risks and responsibilities in the contract (IfM, IBM, 2007). The shift of risks demands a clear understanding of the implications to the business. Techniques described in various chapters of this book provide the initial set of tools that can be used to perform such evaluations. Further development of these tools into more

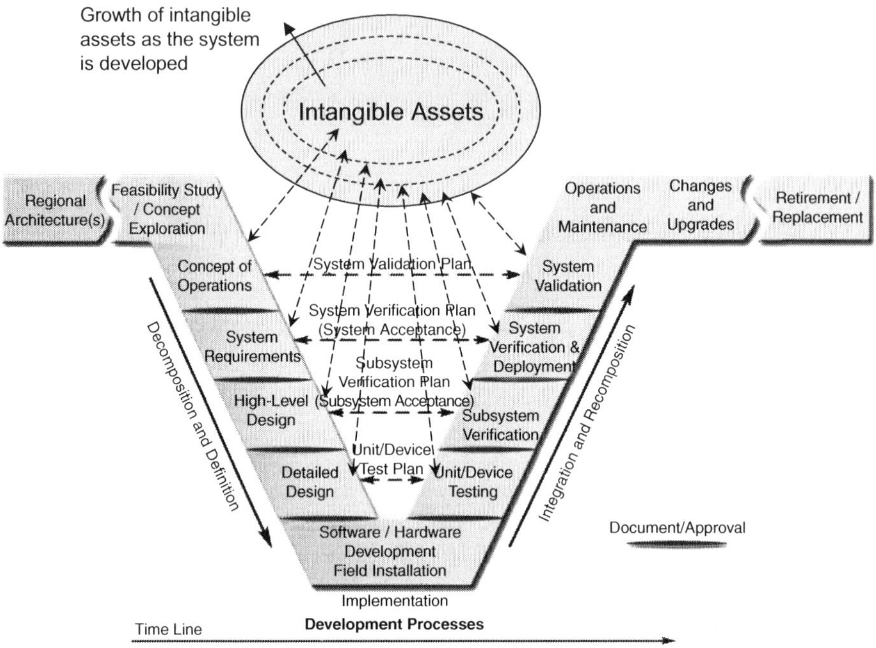

Figure 13.1-1 Intangible assets and other engineering developments

sophisticated and accurate assessment capability is required both as a continuous learning process and as an analytical and investigative research.

Reference

IfM, IBM, 2007. Succeeding through service innovation: a discussion paper. University of Cambridge, UK. p. 33, ISBN: 978-1-902546-59-8.

Index

Edwards Brothers Malloy
Ann Arbor MI. USA
January 4, 2016